Journey to the Viral World:
Electron Micrographs of Viruses

Philippe Roingeard

Journey to the Viral World: Electron Micrographs of Viruses

 Springer

Philippe Roingeard
University Hospital
University of Tours
Tours, France

The original submitted manuscript has been translated into English. The translation was done using artificial intelligence. A subsequent revision was performed by the author(s) to further refine the work and to ensure that the translation is appropriate concerning content and scientific correctness. It may, however, read stylistically different from a conventional translation.

ISBN 978-3-031-77994-7   ISBN 978-3-031-77995-4   (eBook)
https://doi.org/10.1007/978-3-031-77995-4

Translation from the French language edition: "Portraits de Virus - Voyage au coeur des cellules" by Philippe Roingeard, © Presses Universitaires François-Rabelais de Tours 2020. Published by Presses Universitaires François-Rabelais de Tours. All Rights Reserved.

This Springer imprint is published by the registered company Springer Nature Switzerland AG
The registered company address is: Gewerbestrasse 11, 6330 Cham, Switzerland

If disposing of this product, please recycle the paper.

Foreword

The interaction between viruses and humans has played a significant role in our evolution. Endogenous retroviruses have gradually, over thousands of years, evolved our genetic heritage; infections by pathogenic viruses have contributed to a selection of individuals possessing an immune defense system adapted to the elimination of these infectious agents. In this context, we live in a paradoxical era. The twentieth century was a century of extraordinary medical progress with vaccination being one of humanity's major successes and has significantly reduced the burden of infectious diseases (for example, smallpox has been declared by the WHO as an eradicated infection). At the same time, we live in a worrying period where the risk of infection remains very present.

Advances in molecular biology have allowed us to trace the path of several of the pathogenic viruses for humans and to show that they have been present for a long time, thousands of years for example for hepatitis B. Other viruses seem to emerge; this is the concept of *emerging and reemerging viruses*. Yellow fever seemed contained by effective vaccination but recent epidemics have shown how threatening it remains when a vaccination policy is not properly implemented. Viruses like Ebola, Zika or Chikungunya were also very likely present for a long time but have triggered epidemics that led to their discovery. Finally, genetic recombinations from animal viruses have allowed and still allow these viruses to adapt to humans and therefore to infect them. The vast majority of epidemics that we currently observe are indeed due to zoonoses, that is, viruses that infect the animal species.

Of course, we can recall that many infections over the last centuries have very likely gone unnoticed and that the increase in epidemics and pandemics that we observe is in part due, paradoxically, to the progress of public health surveillance; however, we are clearly witnessing a resurgence of these epidemics and the one recently due to SARS-CoV-2 only confirms it. Humans have profoundly modified their ecosystem (and continue to); urbanization, deforestation, human migrations and climate changes create unfortunately ideal conditions for existing viruses to be transmitted to populations suddenly much more exposed and non-immunized, and for animal viruses to infect humans after consumption of food infected by these viruses (infection which, as indicated previously, is generally due to a reorganization of their genetic heritage after "passage" in the animal). This evolution of our ecosystem is unfortunately not the only one to blame; the refusal of vaccination by certain groups, the "fake news" have led to lowering the guard on diseases like measles for which we have an effective vaccination; hundreds of thousands of people die each year from the flu for which a partially effective vaccination also exists. This evolution of our society is clearly terribly worrying and in fact represents a threat to our future.

How to cope? The need for effective research in virology has never been more evident; yet, paradoxically, the importance of this discipline has unfortunately been underestimated over the last fifty years. In fact, we need to recreate a larger community of young researchers in virology to contain and prevent future infections. Within virology, molecular biology has logically asserted itself as the driving element of research, as well as diagnosis and therapeutic research. However, "classic" virology, the art of observing viruses and their culture, is absolutely necessary for the success of research.

It is in this context that the book written by Philippe Roingeard offers a very original and beautifully documented perspective on the importance of direct observation of viruses, their structure, and their morphology (which, by the way, led to the nomenclature; the term "coronavirus" comes from the "spikes" described on their surface). Electron microscopy has played an essential role in the progress in virology and has unfortunately also been largely underestimated in recent years. Let's not forget that it was through electron microscopy that viruses like HIV and hepatitis B virus were first identified!

Philippe Roingeard's book offers us for each virus both a concise and very informative summary of the knowledge we have and spectacular illustrations that really help us better understand who these enemies are. This superb work, based on Philippe Roingeard's remarkable skills in this field, will, I am convinced, be a reference in virology; it will also contribute to the dissemination of knowledge to the general public, a crucial point for populations to appropriate the progress of science.

Christian Bréchot

Professor at the University of South Florida (Tampa, USA), former Director General of The French National Institute of Health and Medical Research (INSERM) and former Director of the Pasteur Institute of Paris, then President of the Global Virus Network (now acting as Vice-Chair of the Board).

Introduction

Since my early days in research, in the mid-1980s, I have spent thousands of hours observing viruses under an electron microscope and have taken thousands of photographs of these viruses. I have always been fascinated by viral structures, their appearance inside or outside infected cells. Over all these years, I have meticulously preserved the most representative of all these photographs, with the idea of creating an educational book. As I progress in my career, I spend more time in the laboratory as a team manager than as a researcher. But I have also greatly enjoyed supervising research work, and some of the photographs in this book were thus taken by my collaborators during projects that we have developed together. I am deeply grateful to them.

With this book, I wanted to apply to virology this famous adage: "a picture is worth a thousand words". So this is not a classic virology book, but rather a "picture book" to help understand what viruses are, their mode of operation. It will only give to the readers a few notions about a complex discipline, but these notions will hopefully encourage them to go further in understanding the world of viruses, with more specialized books. This book does not categorize viruses by family or by pathology, but rather according to their behavior in cells. It is thus more the vision of a cellular biologist and microscopist on viruses than that of a virologist. These images will nevertheless be an opportunity to mention the pathologies induced by the viruses, including those that have attracted particular media attention (HIV, H1N1 flu, Zika and dengue viruses,...), not forgetting of course the recent Covid-19 epidemic.

I wanted to accompany these images with captions that are as simple as possible so that this book can be understood by everyone. Electron microscopy images are, by the principle of the technique, always generated in black and white. Most have however been colorized in this book to make them more attractive or to facilitate the understanding of the subject. This is a "compromise" to the current trend. For a purist, the electron microscopy photograph should remain in black and white since the colors attributed to structures during these stages of artificial colorization have no scientific meaning. However, I have noticed over the years that color photographs were always more successful with the media, even if scientifically they were not the most relevant. So I sought a compromise.

I hope the readers will enjoy this book as much as I have enjoyed observing viruses with my electron microscopes. When studying potentially infected cells or when searching for viruses in a biological fluid, it sometimes takes hours of observation before finding a viral particle. When this is the case, there is a real pleasure in tracking down the virus or viruses, like a detective unmasking a criminal. When it comes to studying massively infected cells, the viruses are then easy to find, and it is often fascinating to observe how they can hijack the activity of the cell to their advantage and implement effective strategies to multiply in very large numbers. This sometimes results in the formation of structures with a remarkable organization that generates spectacular images, some of which are part of this selection.

The research work that I have been able to carry out during my career as a researcher has always been a team effort in the INSERM-University U1259 research unit in Tours (Loire Valley, France), and in the electron microscopy facility of the University and University Hospital of Tours, and I thank all the people who work or have worked alongside me during all these years. Finally, I would like to thank all my colleagues with whom I have developed collaborative research projects, in France or abroad, and especially colleagues from the National Association for Research on AIDS, viral hepatitis and emerging infectious diseases (ANRS-MIE), from the French Society of Virology (SFV) and the National Association of Teachers in Cell Biology from medical schools (ANEBC).

Philippe Roingeard, March 2024

Photo credits (other than those I have taken myself): Julien Burlaud-Gaillard (panels 13, 23, 36, 39, 40, 41, 55); Jean-Christophe Meunier & Eric Piver (planel 51); Pierre-Ivan Raynal (panel 16); Pierre-Yves Sizaret (panels 7, 9, 14); Sonia Georgault (panel 64); Patrice Latron (conclusion).

Glossary

Arbovirus. Virus transmitted by biting arthropods (mosquitoes, ticks).

Capsid/Capsomer. A capsid is made up of many identical protein units (the capsomers) that group together to form a compact structure that surrounds and protects the viral genome.

DNA/RNA. Polymer of nucleotides carrying genetic information, in a double-strand form (DNA) or single-strand form (RNA). In a cell, DNA is transcribed into RNA, which is used to synthesize proteins.

Encephalopathy. Disease affecting the brain, often of infectious origin.

Eukaryote/Prokaryote. Eukaryotes include organisms, unicellular or multicellular, whose cells have a nucleus and organelles delimited by a membrane. Prokaryotes are living beings whose cellular structure does not include a nucleus and organelles, notably including simple unicellular microorganisms, bacteria.

Genome. The entire genetic material of an organism contained in its DNA, except in the case of viruses for which the genome can be either DNA or RNA.

Helicoid/helical. Helix-shaped structure. Some RNA virus capsids form a helical structure, by attaching multiple capsomers to the viral genome.

Icosahedron/icosahedral. Three-dimensional structure of the polyhedron family, containing twenty faces. Some viral capsids have an icosahedral structure, with the viral genome placed inside the icosahedron.

Immunodeficiency/Immunosuppression. State in which a person's immune system is weakened (immunodeficiency), or even inhibited (immunosuppression). Immunosuppression can be deliberately induced to prevent organ transplant rejection.

Oncogene. Gene whose expression induces the development of cancer.

Pandemic. Epidemic affecting a large part of the world's population.

Pathogenicity. The ability of an infectious agent to induce the clinical signs of a disease.

Retrovirus / reverse transcriptase. Virus whose genome is made of RNA but whose infectious cycle includes a step of reverse transcription during which DNA is generated from this RNA. The viral enzyme responsible for this mechanism is called reverse transcriptase.

Syncytium (syncytia in plural). Large cell formed by the fusion of several cells.

Electron Microscopy

Panel 1. With the exception of a few rare «giant» viruses, viruses have a size between 1/80000 th to 1/5000 th of a millimeter. Only electron microscopy allows for sufficient magnification to observe them. Unlike optical microscopy, where the photons are the elementary particles of light that generate an image, electron microscopy uses electrons. When I started my career, we still used special negative films, sensitive to electrons, which had to be developed in a darkroom to make paper prints. In the early 1990s, cameras detecting electrons replaced these time-consuming methods, allowing for much faster generation of digital-type images. These images show the different ways of observing viruses in electron microscopy (on the left, the original images; on the right, these same images colorized by computer. They represent the same virus, the human **immunodeficiency** virus (HIV, the AIDS virus, see also panels 28 to 32), which is about 1/10000 th of a millimeter in size, studied by these different approaches.

Transmission Electron Microscope (TEM)

This type of microscope analyzes the «transmitted» electrons (explaining its name) through a sample, and this is the contrast between the transmitted electrons and those retained by the sample that forms the image. The electrons that pass through the sample imprint the negative film or are detected by the camera, forming a clear area. Those that are retained by the sample leave the area dark, defining an «electron-dense» area. All shades of gray are possible depending on the density of material in the sample. The image at the top shows a cell section. The virus is assembled by budding on the cell surface (in green). This is the **capsid** of the virus (in red) that self-assembles on the surface of the cell to generate this budding. This **capsid** represents a kind of small box that contains the **genome** (the genetic material) of the virus. When budding, the virus surrounds itself with a thin layer of membrane (its envelope, in yellow) borrowed from the cell from which it extracts itself. The image in the center shows these same virus, but in solution. This is referred to as analysis by negative staining. A heavy metal salt is used to obtain a good contrast (negative staining is indeed a misnomer, as it is not a stain but rather a contrasting agent). This contrasting agent concentrates around the virus, which therefore appears in negative (explaining the term «negative staining»). It often happens that the contrasting agent penetrates into the viral particle, allowing to observe different elements inside the virus (as on this photograph where the viral **capsid**, in red, can be distinguished, surrounded by the viral envelope, in yellow).

Scanning Electron Microscope (SEM)

The SEM relies on the use of an electron beam that scans the sample (explaining, here also, its name), and analyzes the electrons that are re-emitted by this sample. Unlike the image at the top which showed a cell section, the image at the bottom shows an entire cell and the SEM allows to visualize the surface of the cell in 3 dimensions. Thus, only the surface of the cell (in green) and the surface of the viruses (in yellow) that come out of this cell can be visualized by this method.

© The University of Tours 2025
P. Roingeard, *Journey to the Viral World: Electron Micrographs of Viruses*,
https://doi.org/10.1007/978-3-031-77995-4_1

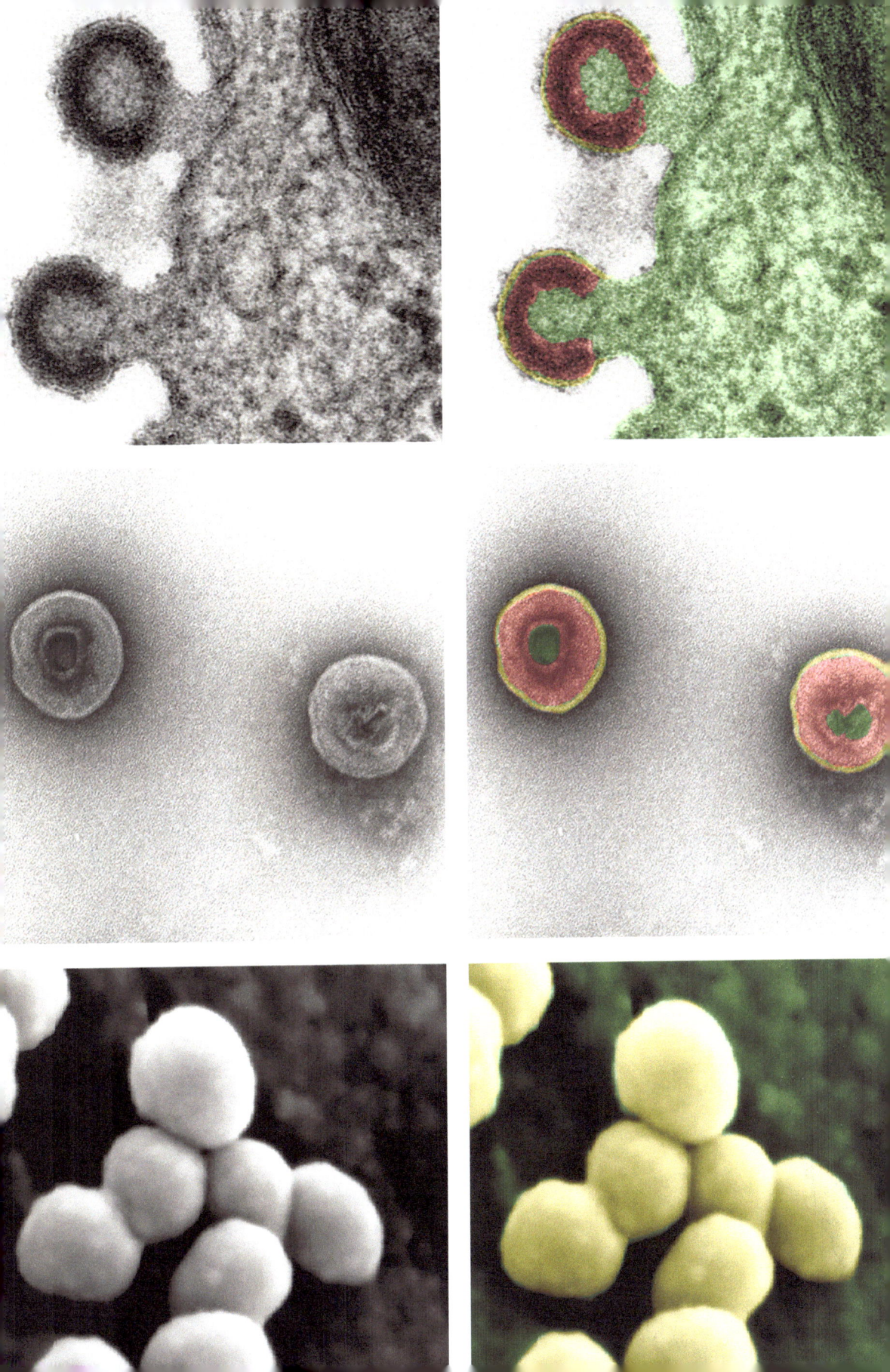

Panel 2. Viruses and bacteria are the two main players in the world of microbes, which can be defined as microorganisms responsible for infectious diseases (knowing that not all microorganisms are pathogens). Diseases as diverse as plague, cholera, syphilis or tuberculosis are caused by bacteria. Diseases like measles, rubella, chickenpox or AIDS are on the other hand caused by viruses. But what are the differences between these two main types of microbes?

These juxtaposed images obtained by scanning electron microscopy show the difference in size between bacteria and viruses. The two viruses (in yellow) are HIV, the AIDS virus. They are spherical particles about 1/10000th of a millimeter in diameter. Some bacteria can also have spherical shapes, but they are much larger. They are indeed perfectly visible with optical microscopes. The bacteria shown in this image (in green) form rod-like structures about 1/2000th of a millimeter in diameter and 1/500th of a millimeter long. Other types of bacteria take the shape of commas or spirals. Regardless of the shape of the bacteria, viruses are therefore much smaller than bacteria.

Viruses
and
bacteria

But this difference in size is not the essential element. Bacteria or viruses are fundamentally different in that bacteria are cells (called «**prokaryotes**»), autonomous in dividing and therefore multiplying, while viruses cannot multiply on their own. To do this, they must necessarily enter a cell and use the machinery of this cell to multiply. They are defined as «obligatory parasites» of the cell. In this sense, viruses are not quite living beings since they are incapable of «living» on their own. Some viruses, called «bacteriophages» (see panel 16), can infect bacteria. Other viruses are capable of infecting **eukaryotic** cells, such as animal cells, plant cells or insect cells. This book will be more specifically devoted to animal viruses, and more particularly to viruses infecting humans.

This fundamental difference explains why drugs against bacteria, the antibiotics, have no effect on viruses. These antibiotics are designed to be specific to the machinery of **prokaryotic** cells (thus bacteria) and to have no effect on **eukaryotic** cells. Otherwise, they would indeed be very toxic to the cells of organisms treated with these antibiotics. Viruses, which are therefore obligatory parasites of a cell, use all the machinery of the cell they infect. They can only be treated by antiviral molecules, which are more or less efficient depending on the different types of viruses. For some viruses, there is no efficient treatment.

New types of viruses have been discovered recently, leading to the concept of «giant» viruses. Isolated in varied environments (freshwater ponds, seawater, permafrost), these viruses capable of infecting unicellular **eukaryotes** like amoebas are almost the size of a bacterium. However, they clearly have the properties of viruses, notably this inability to multiply outside the cell they infect.

© The University of Tours 2025
P. Roingeard, *Journey to the Viral World: Electron Micrographs of Viruses*,
https://doi.org/10.1007/978-3-031-77995-4_2

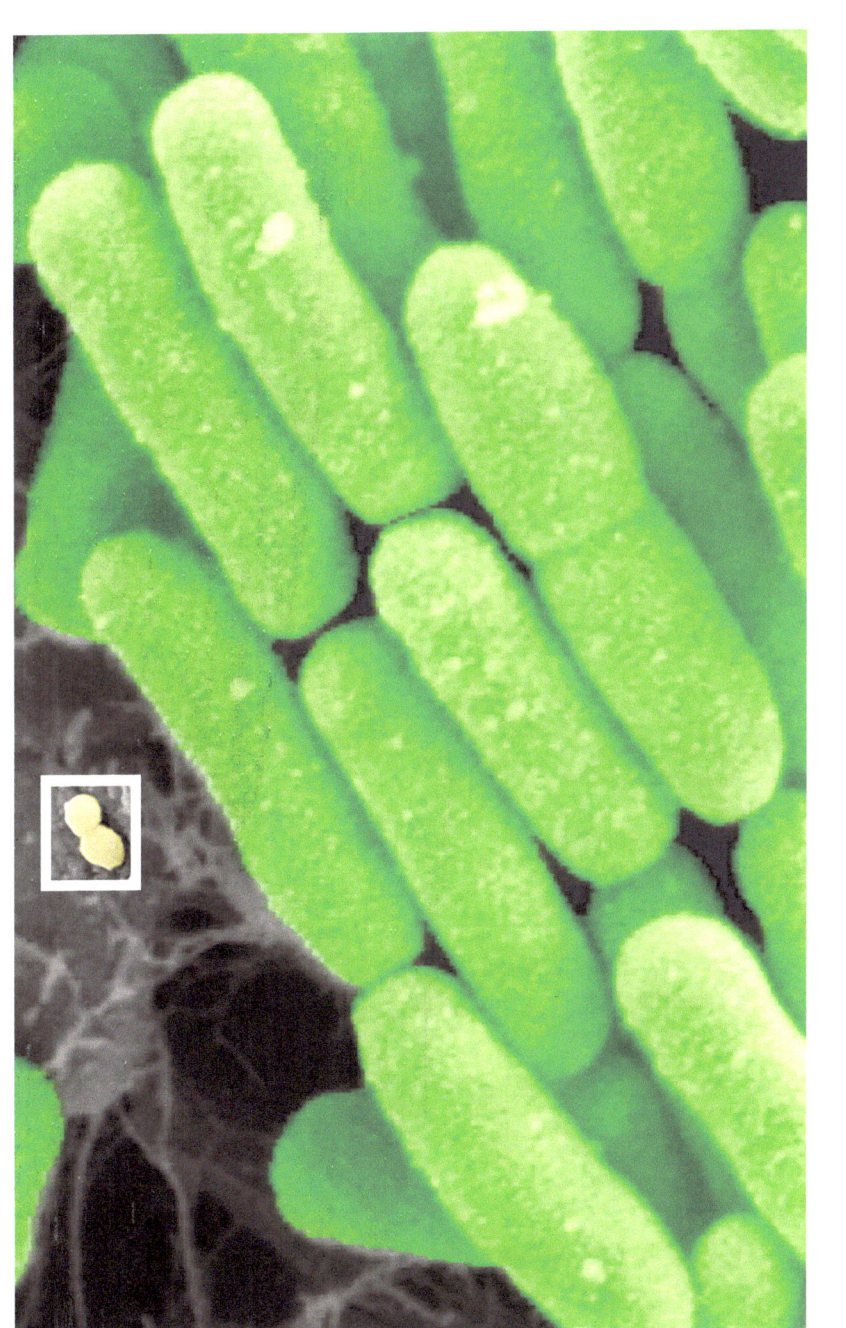

Viral capsid

Panel 3. Every virus has a **capsid**, which can be considered as a box that contains its **genome**, which is made up of a nucleic acid that can be either **RNA** or **DNA**. Some viruses, seen later, also have an envelope. However, the **capsid** is truly the minimal element of the organization of a virus. This **capsid** serves to protect the **genome** of the virus, which carries its genetic information, especially when the virus is in the environment outside the cells. The capsid also serves to deliver the **genome** of the virus into a cell, to initiate infection, and to package the **genome** of a virus that exit an infected cell.

The **capsid** is formed by an assembly of viral proteins that give a well-defined structure to the viral particle. It is often a spherical structure. In some cases, this structure forms **icosahedral** (geometric structure with 20 triangular faces) edifices, as in the case of adenoviruses shown in these images obtained by negative staining. In the bottom photograph, at high magnification, this triangulation and the small spherical proteins that assemble to form these **capsids** can be visualized. These small proteins, basic units for capsid assembly, are called **capsomers.** Each viral particle consists of a **DNA genome** (which is inside the viral particles and is not visible in this photograph), surrounded by a **capsid** made up of 252 **capsomers** organized into 20 triangles. An adenovirus measures about 1/10000 th of a millimeter.

There are many different types of adenoviruses that can infect mammals, including humans. They are responsible for relatively benign infections, such as rhinopharyngitis, conjunctivitis or gastroenteritis. They are quickly eliminated by the immune system. However, the immunity induced does not fully protect against other types, and reinfections by these other types are therefore possible.

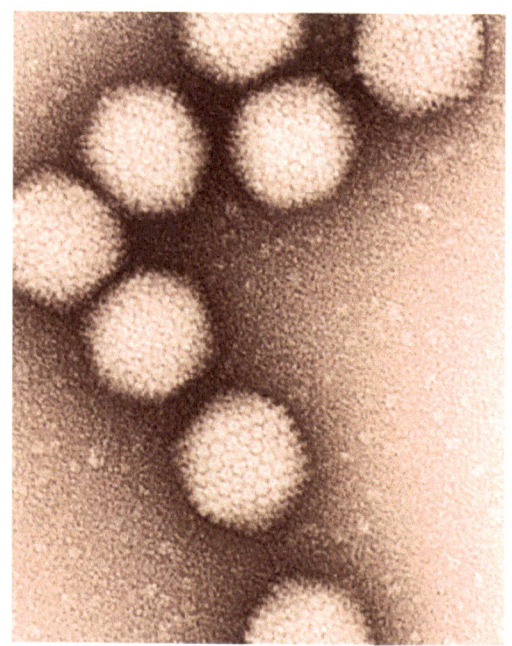

P. Roingeard, *Journey to the Viral World: Electron Micrographs of Viruses*,
https://doi.org/10.1007/978-3-031-77995-4_3

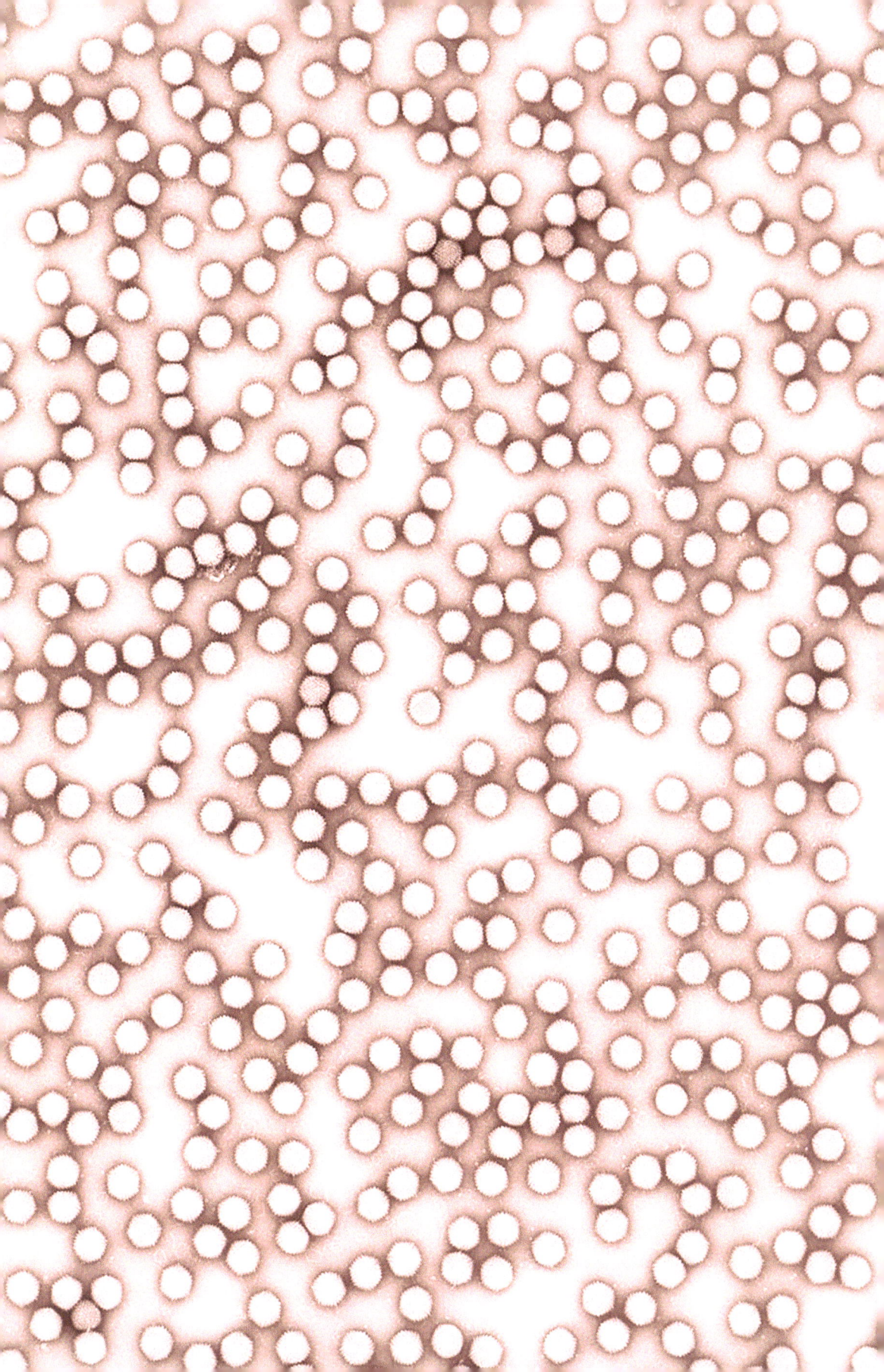

Panel 4. Viruses are "obligate parasites" of a cell. They must necessarily enter a cell to multiply, taking advantage of all the cellular machinery to their multiplication.

These two images, obtained by transmission electron microscopy, show, at two different magnifications, sections of cells infected by an adenovirus. Each dark or grey spherical particle represents an adenovirus, visualized also in section. These images illustrate the extraordinary ability of the virus to multiply at a high level in an infected cell.

Adenoviruses multi- plication

Adenoviruses have a **DNA genome**. When they penetrate a cell, their **capsid** disintegrates to release this **genome**. It takes advantage of the cell's protein synthesis machinery to produce the viral proteins necessary for the assembly of new **capsids**, but also the enzyme allowing this **DNA** to be replicated. These new **genome** copies will be then surrounded with the **capsid** proteins synthesized to generate the new viruses.

As seen in these images, the viral particles accumulate in the infected cell and eventually form structures resembling crystals in which the viruses align, due to their presence in a very high number. By hijacking all the cellular machinery to its advantage, the virus exhausts the cell which can no longer meet its needs and will eventually die. This is indeed the objective of these viruses made up of a simple **capsid** (without an envelope, the so-called "naked" viruses) which have no other way of leaving the cell than to induce the cell burst. The viruses released by the cell burst will then be able to enter other cells to infect them and multiply. In the case of adenoviruses, a single virus that enters a cell can generate at least 10,000 new viruses in this cell in about thirty hours. In theory, each of these new viruses can infect a new cell, which would lead in 3 cycles of infection to produce at least a trillion (one thousand billion) viruses. As previously mentioned, these adenoviruses are nevertheless relatively harmless. They can infect cells from the respiratory and intestinal tracts but they are quickly neutralized by the immune system and the infections caused are ultimately relatively benign.

© The University of Tours 2025
P. Roingeard, *Journey to the Viral World: Electron Micrographs of Viruses*,
https://doi.org/10.1007/978-3-031-77995-4_4

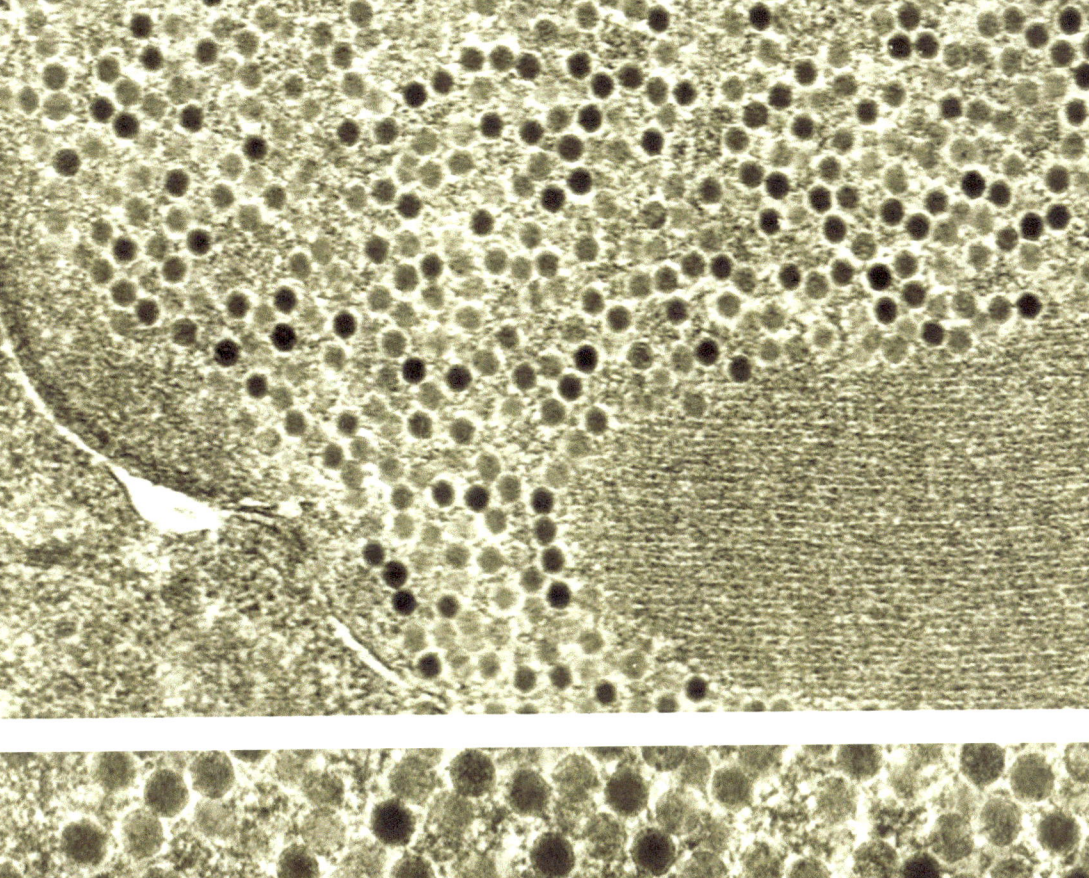

Panel 5. When they are in large quantities, adenoviruses can group together and form ordered structures, in cells or in solution. Here, with this image of negative staining that has been colorized to enhance this trait, the adenoviruses in solution seem to form a mosaic.

Adenoviruses have been used for «gene therapy» protocols, which aim to introduce a gene into a cell. In genetic diseases like cystic fibrosis or severe immune deficiencies, an important gene is defective. Gene therapy is then used to correct this problem by re-introducing the normal gene into this cell. As viruses are capable of entering cells to introduce their **genome**, they are formidable tools for

A mosaic of adeno- viruses

gene therapy protocols. Adenoviruses are particularly interesting for gene therapy because they have a large **DNA genome** that can easily be partially replaced by a cellular gene of interest, by genetic engineering.

However, while adenoviruses have been the subject of many trials in gene therapy protocols in human medicine, they are now less used. One of the obstacles to their use is that people to be treated by gene therapy have often already been exposed to adenoviruses, since these are viruses that are very frequently encountered during benign infections. These people have therefore developed antibodies that prevent the adenoviruses from entering the cells. Even if this is not the case, the treated patients generally develop antibodies against adenoviruses during their first use in gene therapy and these antibodies will then block the virus during subsequent uses of these same viruses during the gene therapy protocol, which then become ineffective.

P. Roingeard, *Journey to the Viral World: Electron Micrographs of Viruses*,
https://doi.org/10.1007/978-3-031-77995-4_5

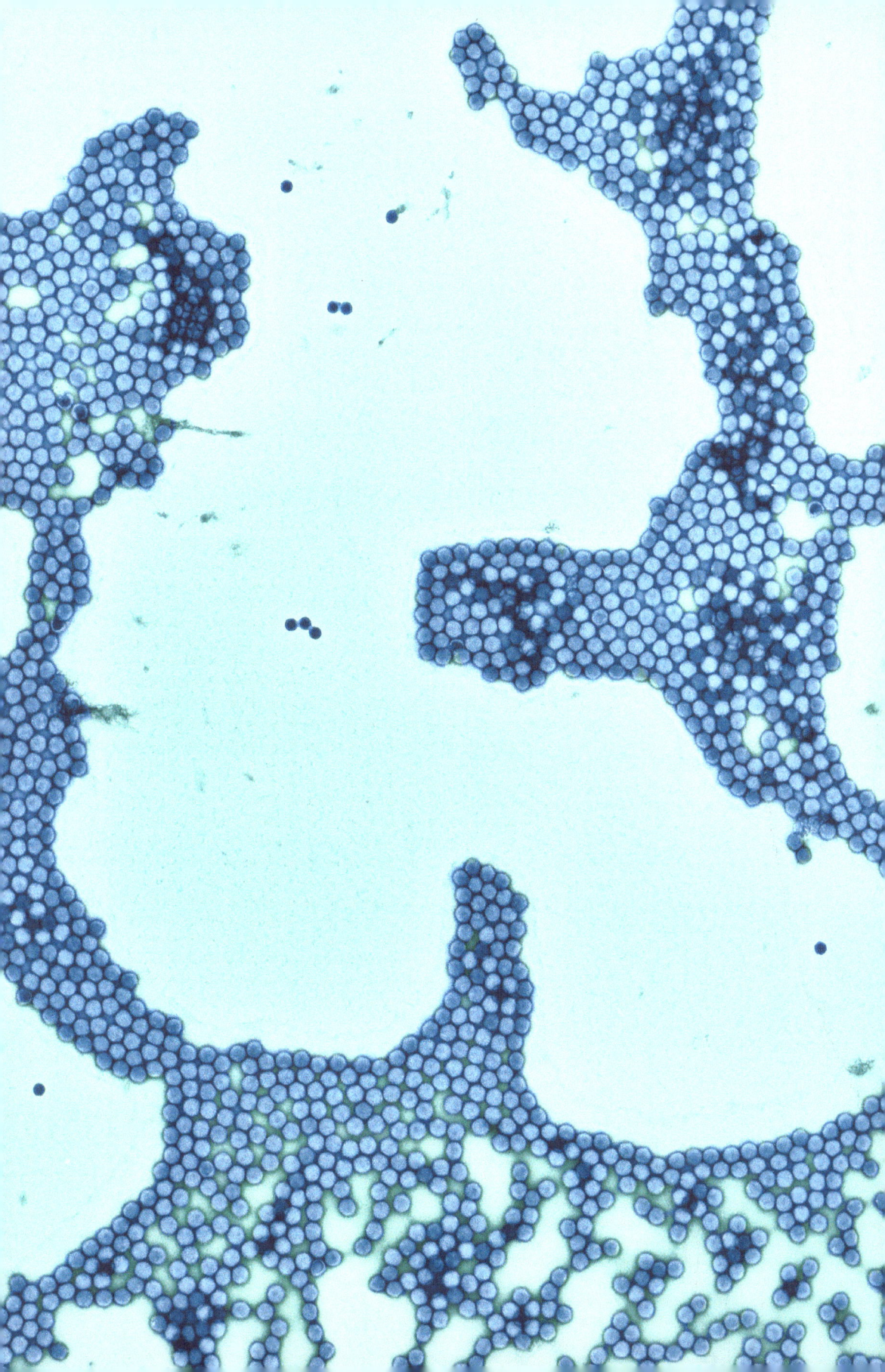

Panel 6. The adenovirus (highly magnified in the top left, see also panels 3 to 5) can help smaller viruses to replicate. These viruses are called AAVs (for Adeno-Associated Virus, highly magnified in the top right). AAVs are tiny viruses (1/80000th of a millimeter) that infect humans and non-human primates. They form small **icosahedrons**, consisting of only 60 protein subunits (the **capsomers**). They are very common in humans and are not attributed to any pathology. They induce a weak immune response and are therefore unnoticed during an infection.

These small viruses are unable to multiply on their own. They require the presence of an adenovirus in the same cell they infect in order to replicate, using certain proteins synthesized by the adenovirus to replicate their **DNA genome**. When this is the case, this mechanism is particularly effective, generating up to a million AAVs per cell. The replication of the AAV then takes over the replication of the adenovirus. It is therefore possible that these small

Adeno-associated virus

viruses, which are still poorly understood, protect us from infections by adenoviruses, or perhaps even other viruses. Indeed, while we have acquired a great deal of knowledge about the viruses that induce pathologies, very little is known about the world of viruses that live or pass transiently in our cells without causing pathologies.

Currently, AAVs are mainly studied to develop gene therapy protocols. Like adenoviruses, they can carry a cellular gene to restore the function of a gene in a human cell containing a defective gene. They have the advantage of being poorly recognized by the immune system and are therefore more effective than adenoviruses.

In the image at low magnification, three type of particles can be observed: adenoviruses, AAVs, and numerous very small particles that represent isolated **capsomers** of the adenoviruses.

© The University of Tours 2025
P. Roingeard, *Journey to the Viral World: Electron Micrographs of Viruses*,
https://doi.org/10.1007/978-3-031-77995-4_6

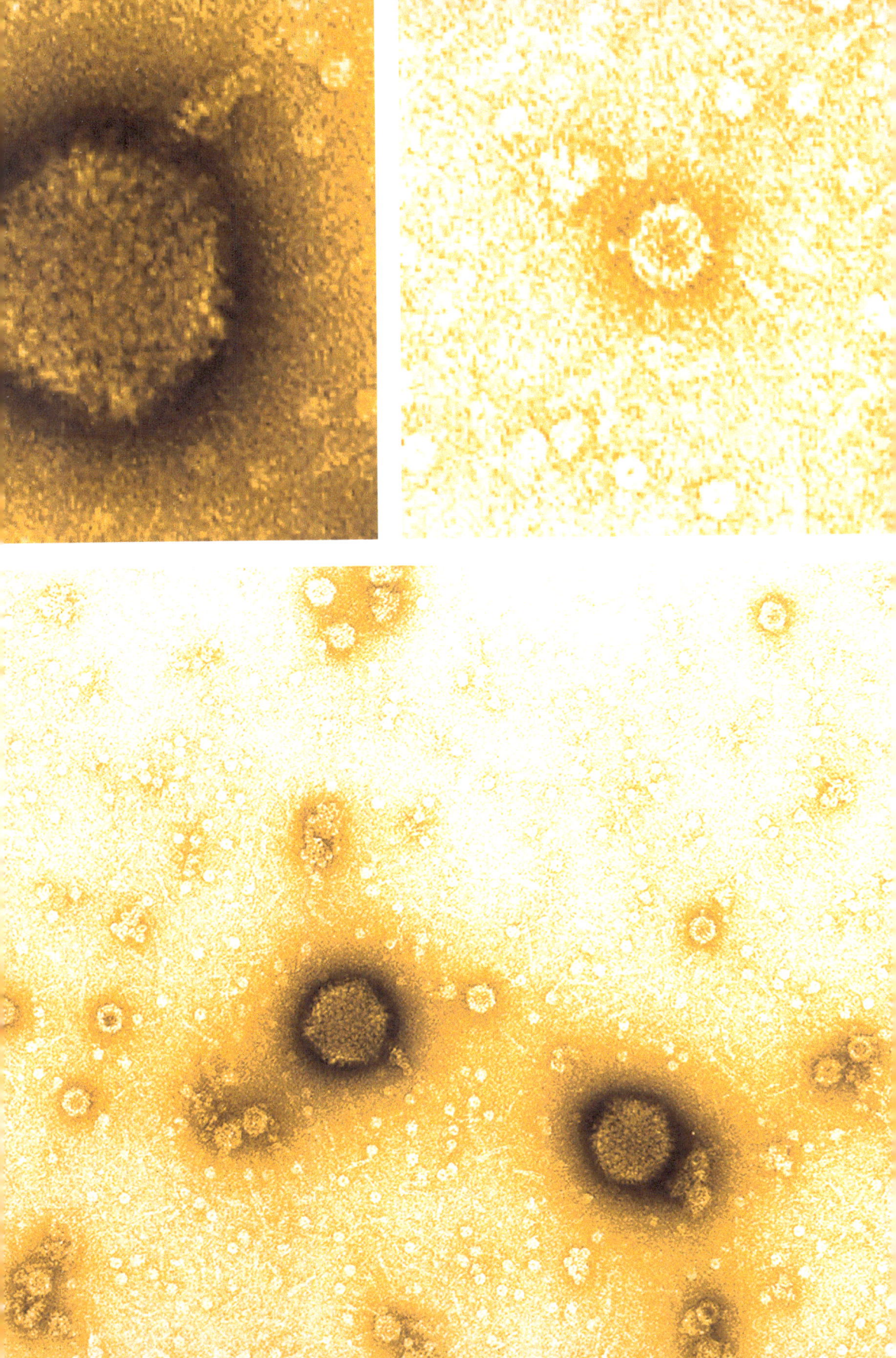

Panel 7. This virus is another example of a small "naked" virus, thus devoid of an envelope. In this photograph, its **capsids**, made up of 72 subunits, the **capsomers**, forming a sphere of about 1/50000 th of a millimeter, are visualized in blue. These **capsids** contain the **genome** of the virus, consisting of a very small **DNA**.

This virus, from the polyomavirus family, is very common in humans and is not associated with diseases, except in people with a weakened immune system. This is particularly the case for transplant patients, especially kidney transplant patients. This virus was first identified in the urine of a kidney transplant patient, whose initials B.K. gave the virus its name. This virus can multiply in kidney cells and is associated with serious diseases in kidney transplant patients, which can lead to graft rejection.

It is often compared to another very similar virus, the JC virus, whose name also comes from the initials of the patient in whom it was isolated, and which has the same natural history as the BK virus. These viruses have existed for a very long time within the human species, with which they have co-evolved. The analysis of the **genomes** of JC viruses isolated from different subjects has thus shown that certain modifications in the **DNA** of the virus are specific to certain ethnic groups. Thanks to these analyses, these JC viruses have been able to constitute real markers of certain human migrations, notably demonstrating that the first humans who settled on the American continent came from Asia.

© The University of Tours 2025
P. Roingeard, *Journey to the Viral World: Electron Micrographs of Viruses*,
https://doi.org/10.1007/978-3-031-77995-4_7

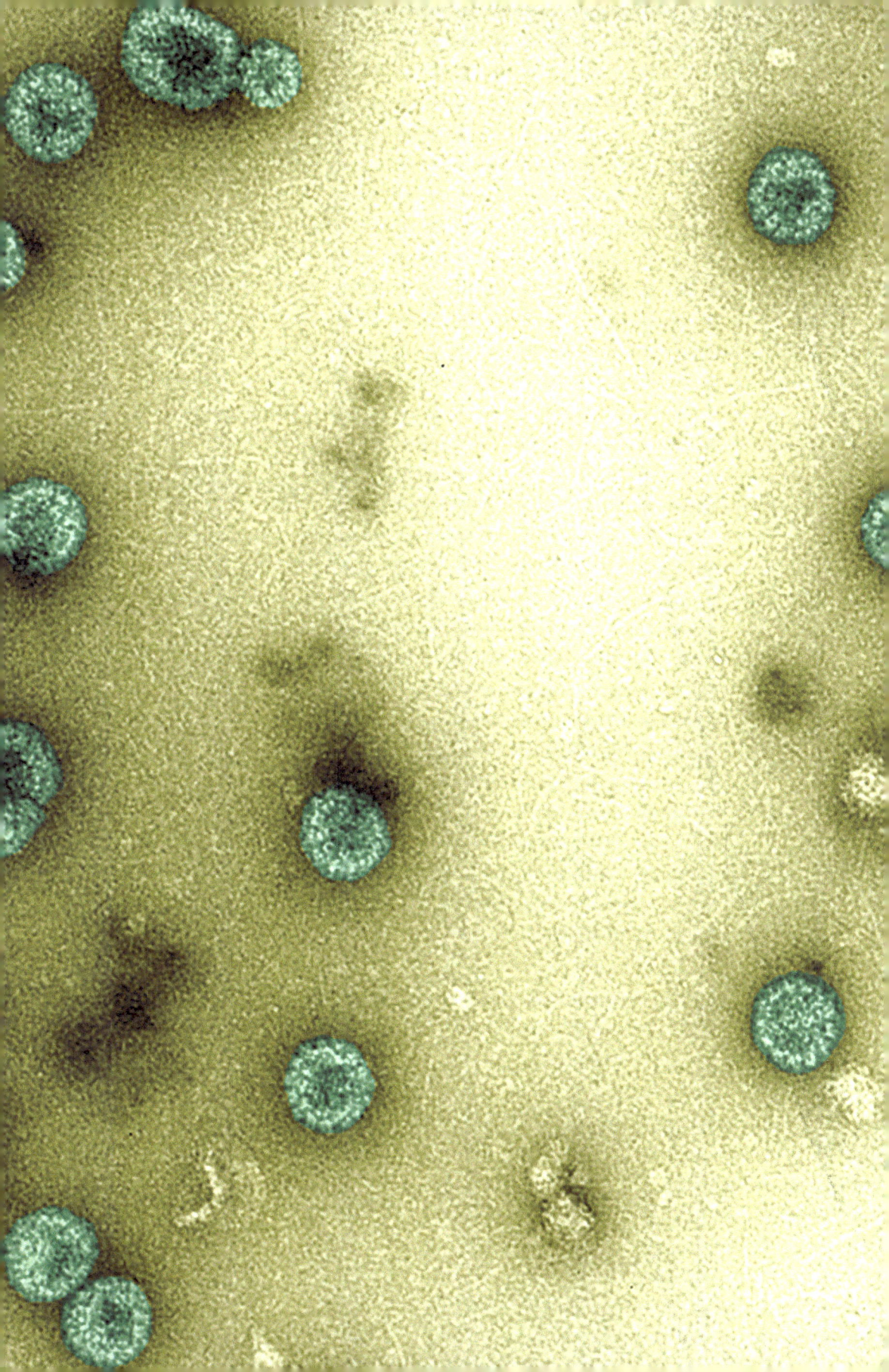

Panel 8. This image shows the intensity of the BK virus multiplication in an infected kidney cell. The viral multiplication is so important that the viral particles that accumulate in large quantities in the cell form structures resembling a crystal (also visualized at high magnification in the photograph below). With such intensity of multiplication, it is clear that the virus hijack all the cellular machinery to its advantage, leading progressively to cell death. This image shows that the cell has lost its internal structure and is dying. The cell burst will allow the viral particles to exit, being then able to infect other cells.

BK virus multiplication

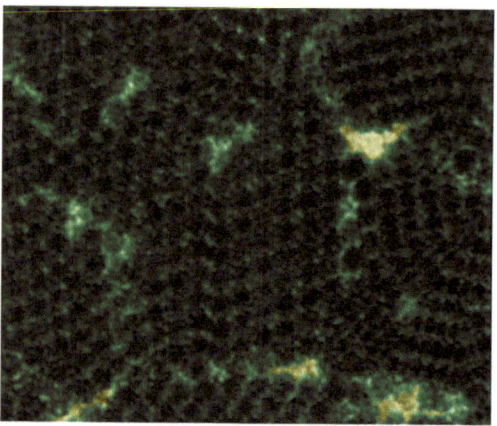

Unlike the adenovirus whose large **DNA genome** allows the synthesis of an enzyme used to replicate the **viral genome**, this type of virus, which has a very small **DNA genome**, does not contain a sequence allowing the synthesis of a replication enzyme for its **genome**. The virus uses a cellular enzyme, the one that duplicates the **DNA** contained in the chromosomes at each cell division, to replicate its own **genome**. Each time the cell divides, it indeed has to duplicate its **genome** using DNA replication enzymes, the **DNA** polymerases. For this mechanism to be as efficient as possible, the virus synthesizes proteins that manipulate the cell cycle so that the infected cell is forced to divide very frequently. The virus thus manipulates the cell cycle to multiply its **genome** at the same time as the cell duplicates its chromosomes. These mechanisms illustrate the ability of viruses to divert cellular machinery to their advantage.

© The University of Tours 2025
P. Roingeard, *Journey to the Viral World: Electron Micrographs of Viruses*,
https://doi.org/10.1007/978-3-031-77995-4_8

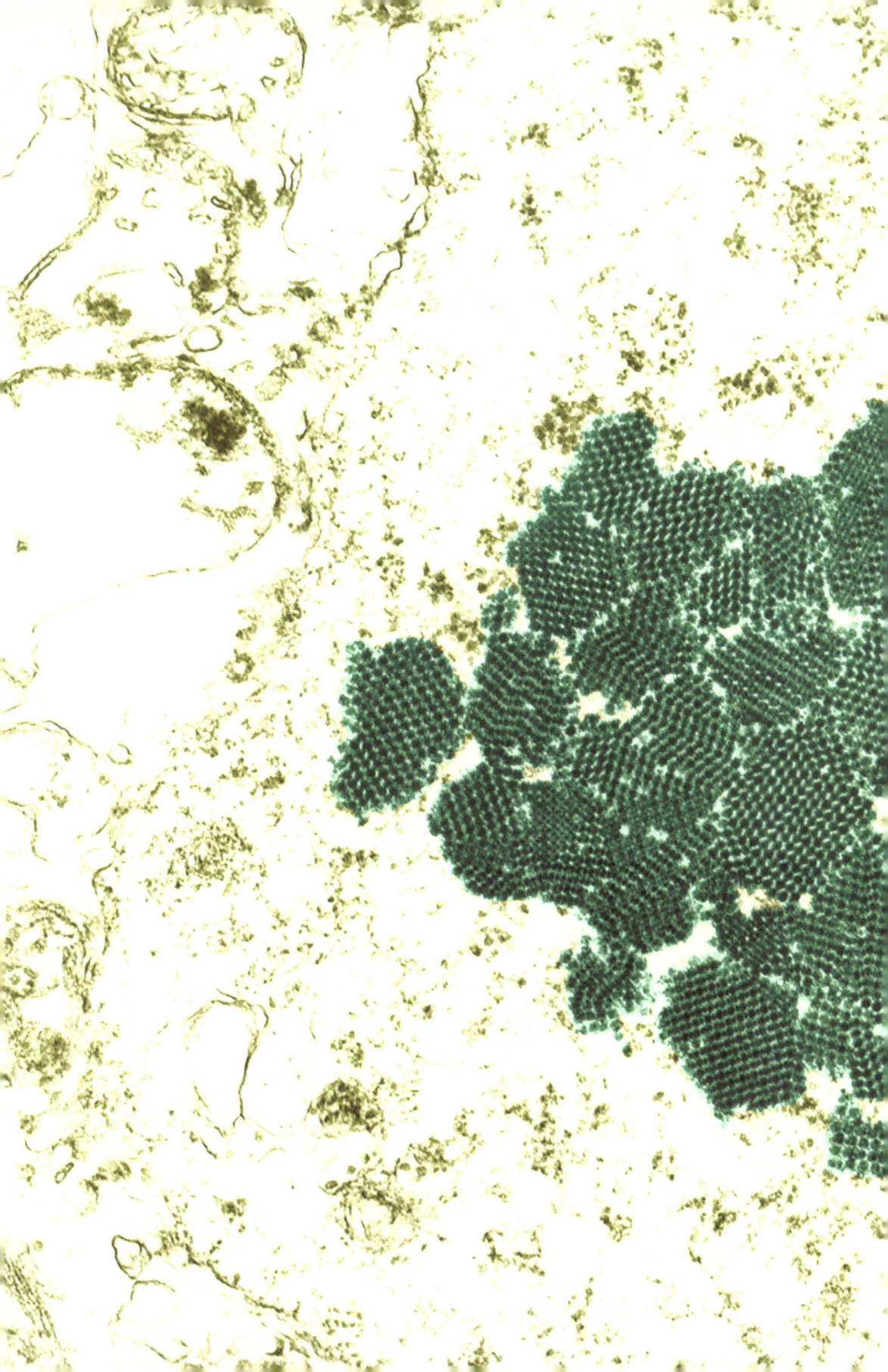

Panel 9. Human papillomaviruses, observed here by negative staining, represent another family of **DNA** viruses with a small **genome** packed into a small spherical **capsid** (about 1/50000 th of a millimeter). But these are much more concerning than the viruses mentioned so far, constituting a real public health problem. While most infections by these viruses give no symptoms, some indeed evolve into cancers.

This family of viruses is very large as there are over 200 different subtypes of human papillomaviruses. Some are transmitted through the skin, others sexually. In some individuals, these viruses induce abnormal cellular proliferations : warts for the viruses that are transmitted through the skin ; small benign tumors called condylomas for those transmitted sexually. These small benign tumors have been known since antiquity as they were already described by Hippocrates. Some subtypes of this large family are more aggressive than others, and make these condylomas can evolve into cervical cancers. This type of cancer is the 4th most important cancer in women worldwide, after breast, lung and colorectal cancer. Since 2005, there has been a vaccine against the subtypes of papillomavirus involved in cervical cancer, recommended for all teenager young women.

Human papilloma-viruses

© The University of Tours 2025
P. Roingeard, *Journey to the Viral World: Electron Micrographs of Viruses*,
https://doi.org/10.1007/978-3-031-77995-4_9

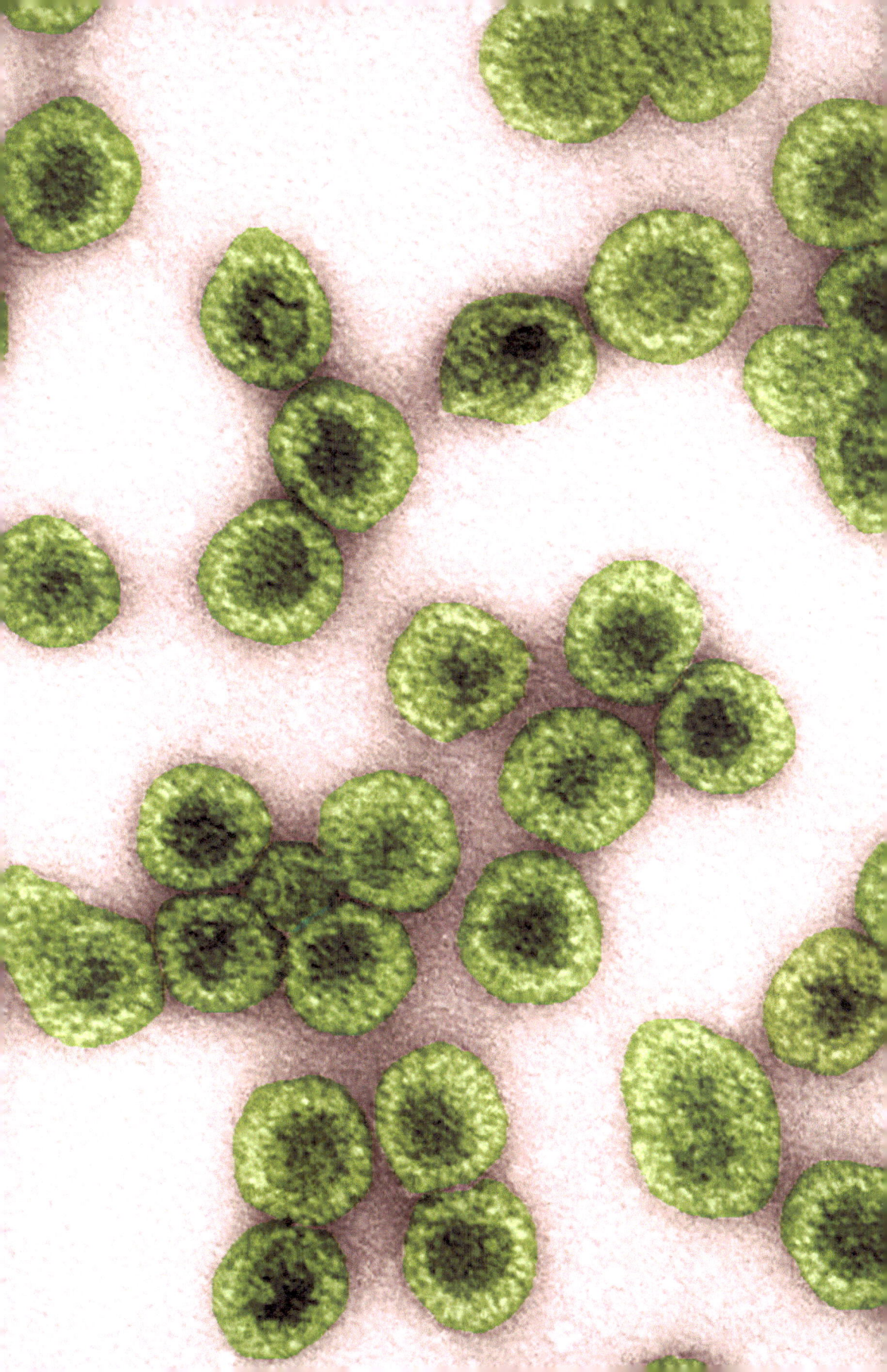

Panel 10. This image of a papillomavirus-infected cell section shows the presence of viral **capsids** formed inside the cell nucleus (in green). It is actually not in the strategy of these viruses to make host cells cancerous; this carcinogenic risk is ultimately "accidental", an event that is not specifically desired by the virus. As for the viruses with a small **DNA genome** seen previously, the interest of the papillomavirus is to multiply very efficiently, then to burts the cell in order to be released from it and to be able to infect other cells. Like the BK virus (panels 7 and 8), the papillomaviruses manipulate the cell cycle to force cells to divide, so that

Human papillomaviruses and cancer

their own **genome** is replicated in addition to the cell's DNA by the cellular replication enzyme. However, when this infectious cycle becomes abortive, these papillomaviruses transform the normal cells into cancerous cells, meaning that the complete infectious cycle which results in cell burst is no longer completed, which leads to cell survival. This happens when the small **DNA genome** of these viruses integrates into the cell's chromosomes. In this case, the cell continues to produce the viral proteins able to manipulate the cell cycle, forcing the cell to divide, without the complete virus being produced and accumulating to burst the cell. These conditions, where cells divide abnormally without any cell death induced by the virus constitute a first step towards the development of a cancer. The cells survive and perform very rapid divisions; they are then subject to instability of their **genome**, which acquires mutations that can lead to cancer.

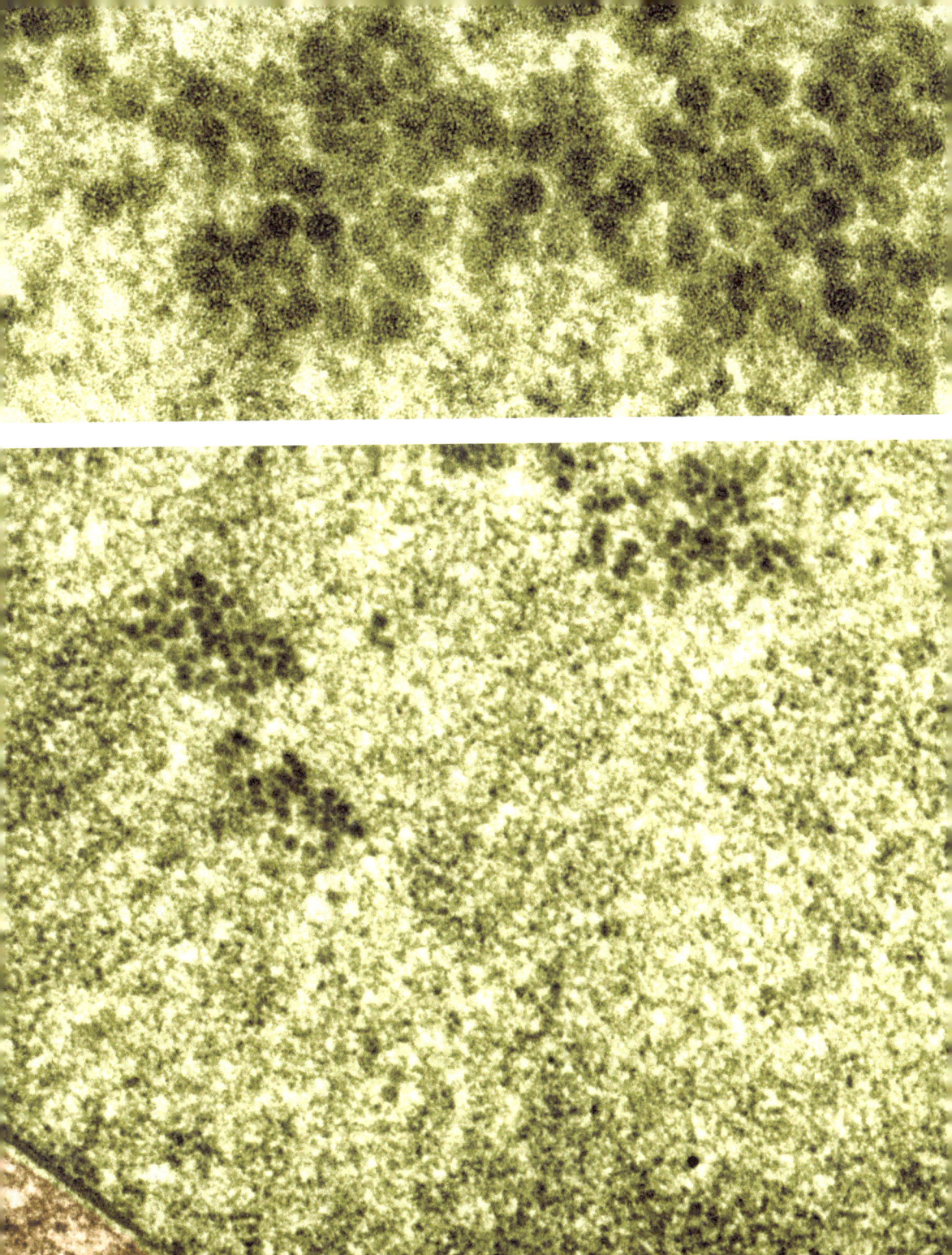

Panel 11. Rotaviruses are also naked viruses, with a **capsid** about 1/10000 th of a millimeter. Their **genome** is made up of several segments of **RNA**, present in this **capsid**. When the virus infects a cell, it hijacks the cellular machinery to make the cell synthesize its own proteins that will constitute this **capsid** as well as the viral enzyme that will replicate the different segments of **RNA** making up the viral **genome**. All these segments (11 in total) are incorporated into each new **capsid** formed. This image shows an infected cell, where an area of the cell (in blue) contains rotaviruses that have just been assembled. The three inset images show isolated rotaviruses, in solution, observed by negative staining. Their structure is very characteristic, their **capsid** being made up of three layers of proteins, giving the viral particle a wheel-like appearance. The virus indeed gets its name from the Latin word *rota*, which means wheel.

Rotavirus

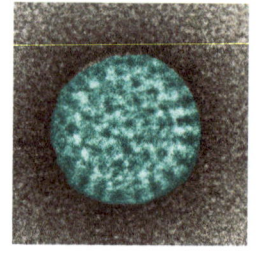
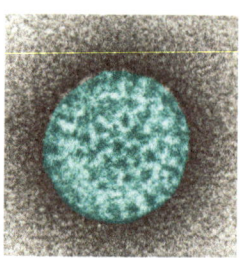
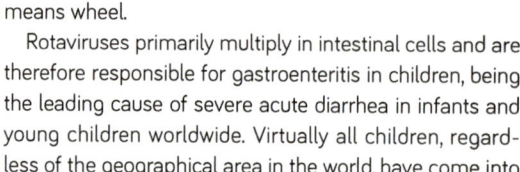
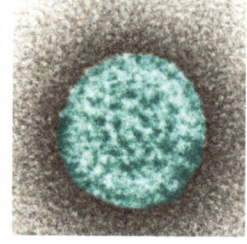

Rotaviruses primarily multiply in intestinal cells and are therefore responsible for gastroenteritis in children, being the leading cause of severe acute diarrhea in infants and young children worldwide. Virtually all children, regardless of the geographical area in the world, have come into contact with this virus within the first 5 years of their life. The pathology is directly linked to the aggression of the intestinal cells in which the virus multiplies. Fortunately, these can regenerate quickly and the immune system can usually eliminate the virus. The main part of these infections therefore trigger few symptoms, but some result in episodes of acute diarrhea. These can be very serious as they lead to dehydration, especially in a context of malnutrition. Human-to-human transmission through water or food contaminated by fecal matter is easy, especially since the virus is present in very large quantities in the stools of a child suffering from gastroenteritis (up to a billion viruses per gram of stool).

© The University of Tours 2025
P. Roingeard, *Journey to the Viral World: Electron Micrographs of Viruses*,
https://doi.org/10.1007/978-3-031-77995-4_10

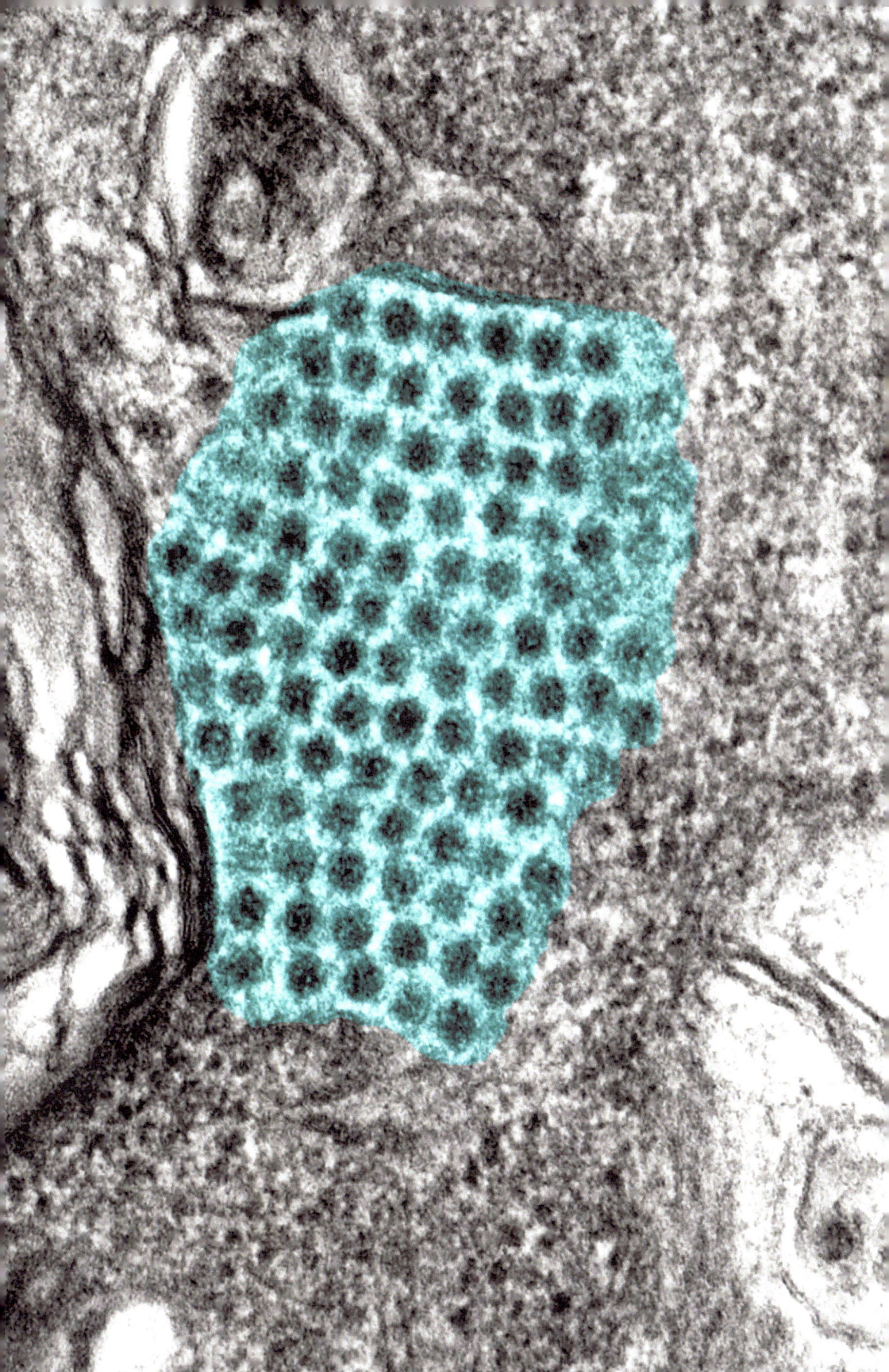

Panel 12. Rotaviruses are part of a very large family of viruses called reoviruses. While the pathology induced by rotaviruses is well known, little is known about the other members of the family. Their name actually comes from the words *Respiratory-Enteric-Orphans* meaning that they can infect cells of the respiratory tract and digestive system, and that they are "orphans" when it comes to attributing a pathology to these viruses. However, we now know that some members of this family, especially rotaviruses, can induce pathologies. But this historical name has nevertheless been preserved.

It seems that in some cases, a few viruses from this family may induce respiratory or neurological pathologies in humans. Theses images at different magnifications show a virus from the family of reoviruses multiplying on a very large scale in human cells, forming structures resembling crystals. This particular

The large family of
reoviruses

reovirus has been identified as responsible for a serious brain diseases, an **encephalopathy**. Like the rotavirus, the other viruses in this family consist of a **capsid** multilayer and a **genome** made of segments of **RNA** (10 to 12 depending on the virus).

The viruses of this large family can infect many vertebrate animal species, but also insects and plants. In principle, there is a species barrier that means that viruses from one species are not transmitted to another. However, some reoviruses cause severe diseases in plants by being transmitted by insects in which they can multiply, but apparently without causing pathology. In humans, respiratory infections have been reported in cases of species barrier crossing from a bat reovirus to humans. But this remains a rare event. The reovirus shown in these photographs was isolated from children of the same family who had been in contact with pigs; a species barrier crossing was therefore suspected.

Reoviruses have recently been identified as agents of serious diseases in fish, causing severe epidemics in salmon farms.

© The University of Tours 2025
P. Roingeard, *Journey to the Viral World: Electron Micrographs of Viruses,*
https://doi.org/10.1007/978-3-031-77995-4_11

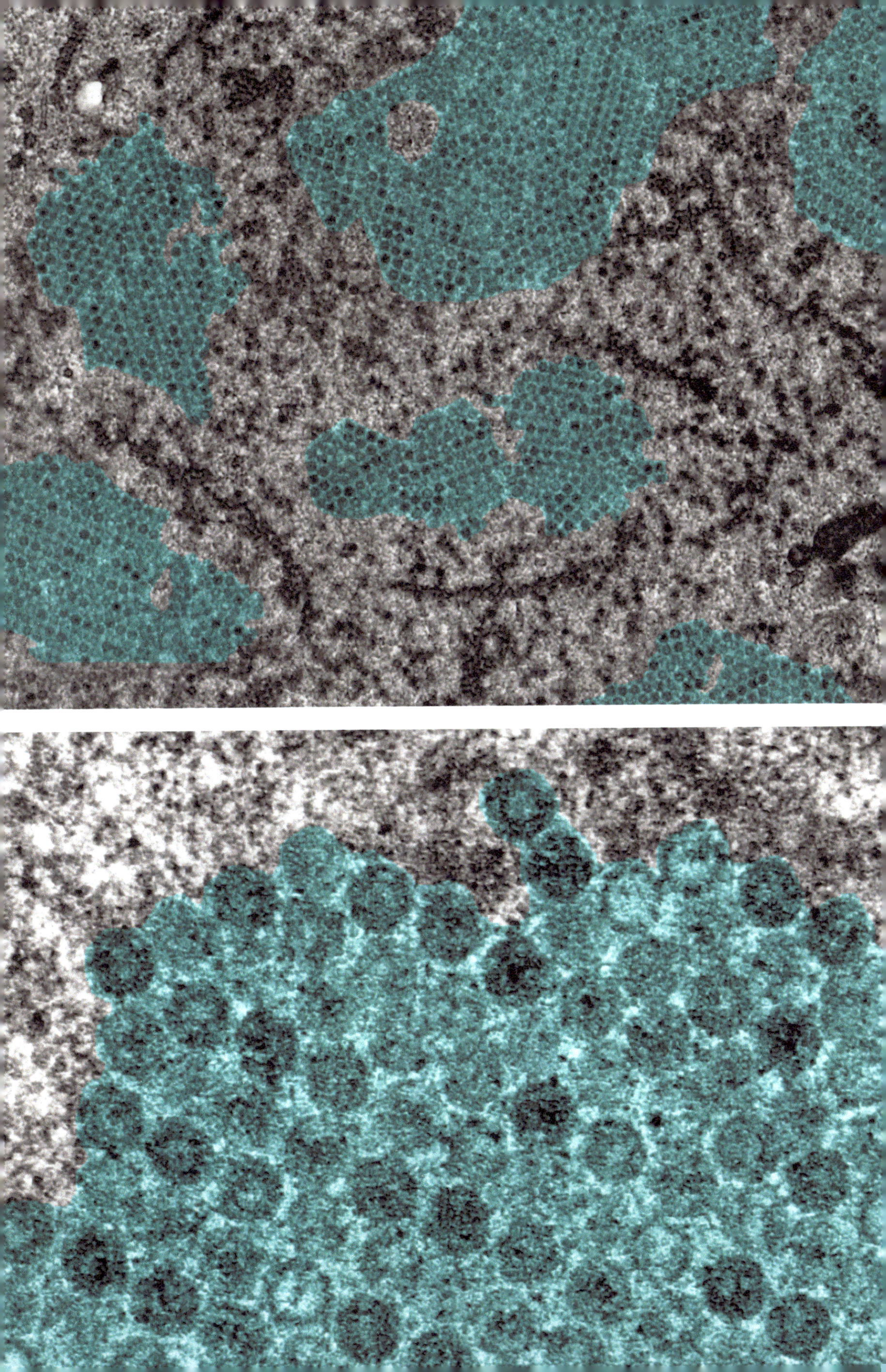

Panel 13. The human parvovirus B19 is a very small virus, 1/80000 th of a millimeter. Its name derives from the Latin word *parvum*, meaning small. Its **capsid** is **icosahedral**, containing a small **DNA genome**. It was discovered by chance, in 1975, in the blood of an infected (but symptomless) subject, and the other part of its name comes from the code of the tube in which it was identified (number 19 in panel B). Pathological significance of the infection was only attributed to this virus a decade after its discovery, when it was found to be responsible for megalerythema epidemicum, a benign childhood rash. Strictly human, it is transmitted mainly via respiratory droplets

Human parvovirus B19

and hands. The virus is highly contagious, with infections mostly occurring in children when they enter school. Infected children rapidly develop a protective immune response against re-infection.

However, it can also affect adults who have never been in contact with the virus before. These infections are usually asymptomatic, or give rise to a mild flu-like syndrome with fever, muscle aches and headaches. In most cases, infected adults may also develop a rash and joint pain. In all cases, these infections remain benign and uncomplicated. Special attention should be paid by non-immune pregnant persons as there is a risk of the virus doing harm to unborn babies especially in late stages of pregnancy.

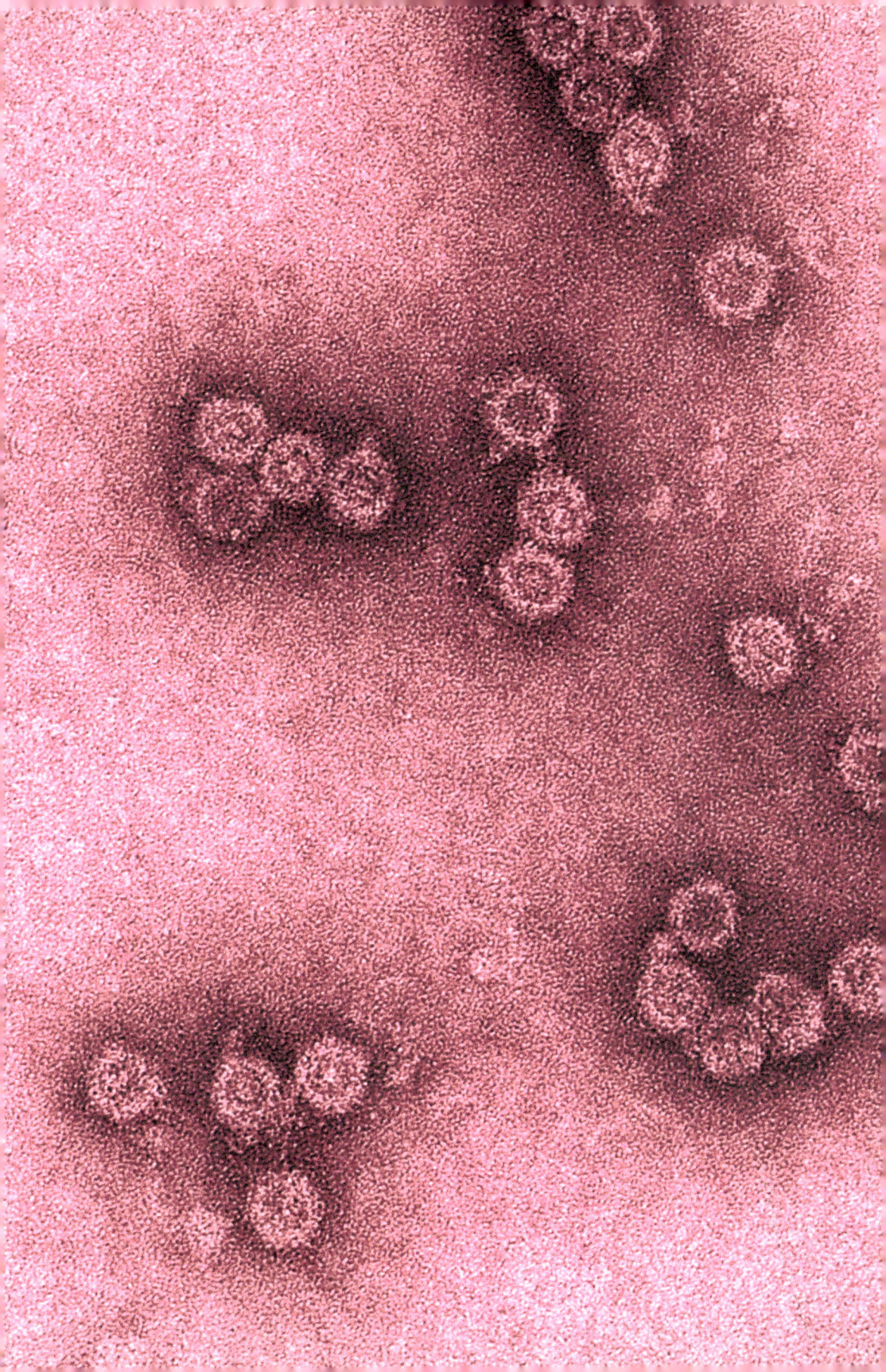

Panel 14. The Tobacco Mosaic Virus (TMV) is the emblematic figure of viruses with **helical** symmetry, its **capsid** forming a helix with a regular pitch around the viral **genome**. It has been the subject of numerous studies, and for a very long time since it was the first identified virus. The disease induced by this virus is called mosaic because it introduces areas of discoloration in the leaves of the plant, which eventually dies. By the end of the 19 th century it was known that the disease was linked to a transmissible agent, but it was not until 1930 that it was identified as a virus.

The tobacco mosaic virus

As these images show, the **helical capsid** of the TMV is remarkably regular. The helix assembly is done by an association of **capsomers** that form a tube around the viral **RNA genome**. These tubes are about 1/3000 th of a millimeter long , 1/80000 th of a millimeter in diameter, with a small channel in the center. These tube-shaped structures are very compact and protect the viral **genome** very well in the environment outside the cells. The virus is therefore very resistant and is easily transmitted from plant to plant. In addition to tobacco, it infects plants such as tomatoes, peppers, cucumbers or even ornamental flowers. It does not, however, infect animal species. This virus has gained renewed interest in recent years for biotechnological approaches. Researchers have shown that a genetically modified TMV can bind to the surface of Lithium battery electrodes and increase the capacities of these batteries.

There are also helical symmetry viruses that infect humans. These are always viruses that have, in addition, an envelope surrounding this **helical capsid** .

© The University of Tours 2025
P. Roingeard, *Journey to the Viral World: Electron Micrographs of Viruses,*
https://doi.org/10.1007/978-3-031-77995-4_12

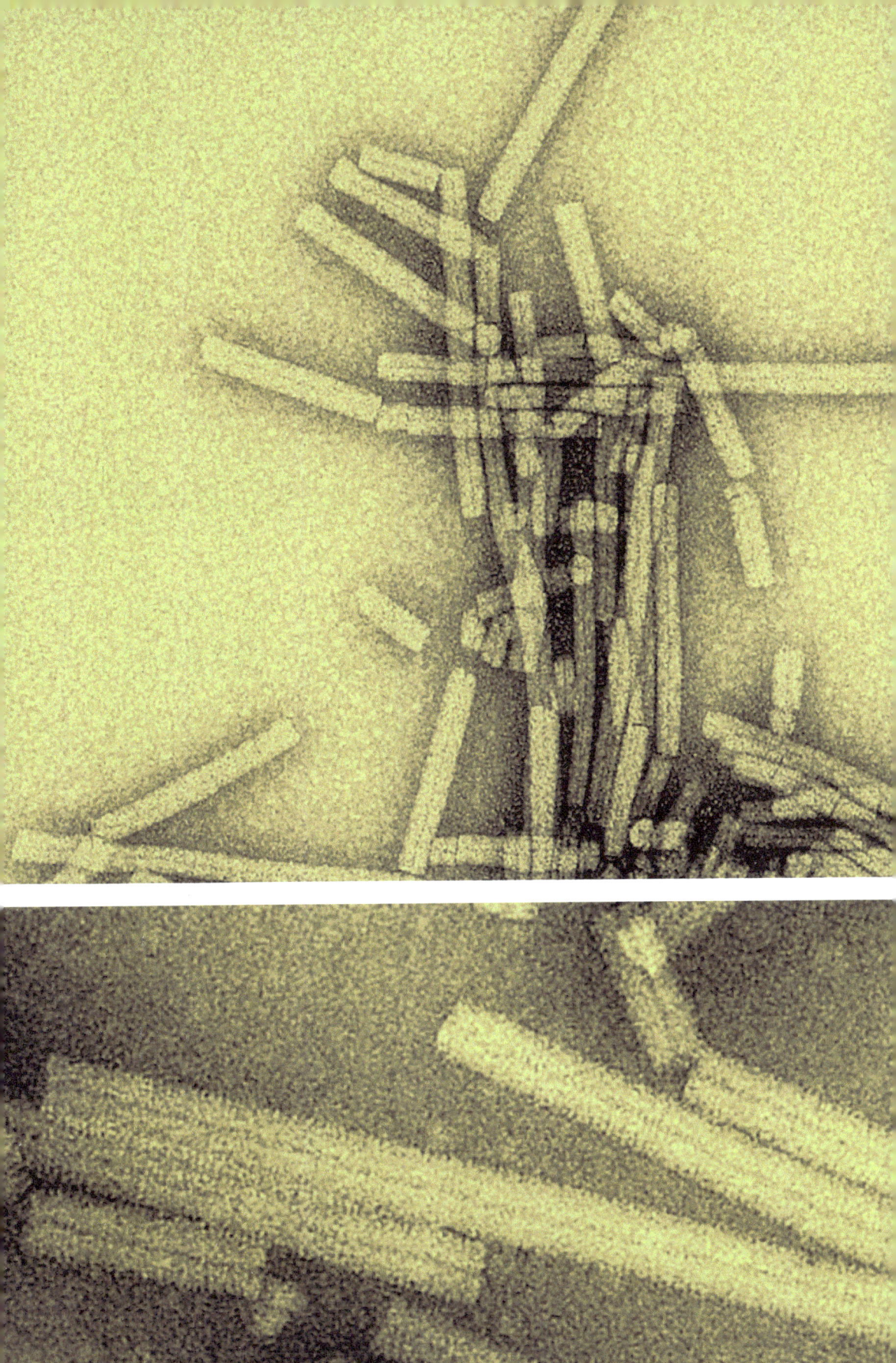

Panel 15. These images show two types of enveloped viruses, for which the **capsid** can be **icosahedral** (like a herpes virus, top left; also see panels 18 to 22) or have **helical** symmetry (like a Respiratory **Syncytial** Virus, or RSV, top right; also see panel 17).

The viral envelope is made by a portion of membrane that a virus borrow from the host cell, at the moment it comes out the cell. This envelope that surrounds the **capsid** of the virus is therefore made up of lipids, like a cell membrane. In addition to lipids, this envelope contains viral proteins, which are called envelope proteins. The crenellated appearance on the surface of the RSV is due to the presence of these envelope proteins.

An enveloped virus is always more fragile than a naked virus in the external environment. This envelope, mainly made up of lipids, is indeed much easier to break than a **capsid**, which is formed of associated proteins giving a very compact structure. A naked virus enters a cell thanks to the presence of a specific receptor on the surface of the cell to which its capsid proteins attach. In the case of an enveloped virus, it

Enveloped viruses

is the envelope proteins that recognize the receptor on the surface of the cell that will be infected. However, if a viral envelope is broken, the **capsid** is released into the external environment but the virus can no longer be infectious without its envelope proteins.

Bellow each image of the complete viruses, an image of the same virus for which the envelope has been experimentally broken is shown, in order to visualize the **capsid**. The image at the bottom left shows the **icosahedral** structure of the herpes virus **capsid**. The image at the bottom right shows the rod-like structure of the RSV **helical capsid**.

© The University of Tours 2025
P. Roingeard, *Journey to the Viral World: Electron Micrographs of Viruses*,
https://doi.org/10.1007/978-3-031-77995-4_13

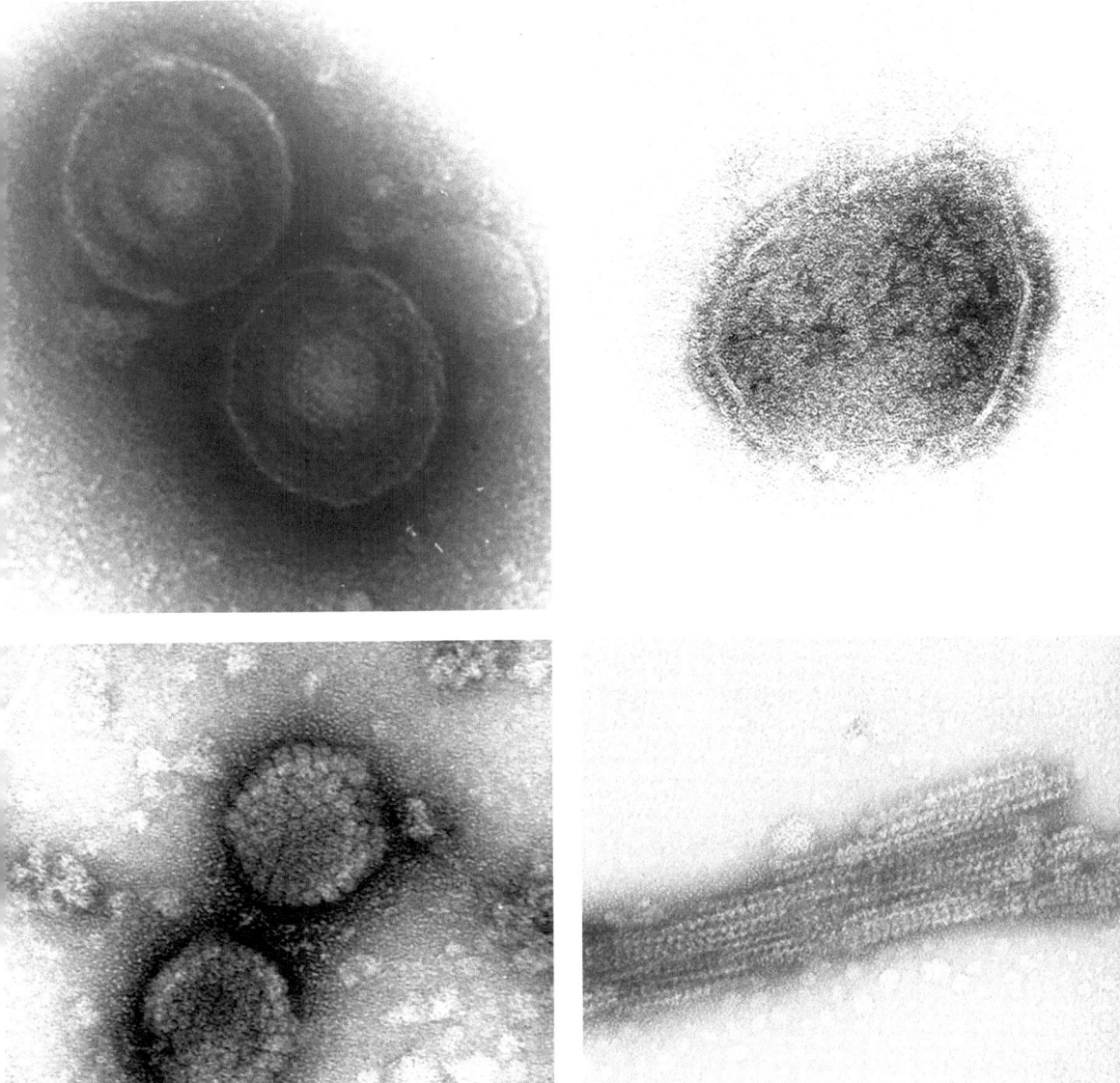

Panel 16. Some viruses have **capsids** with mixed symmetry. They possess an **icosahedral** head and a tail with **helical** symmetry. However, these are quite particular viruses, as they are specific to the bacterial world. These viruses, called bacteriophages (literally "bacteria eaters"), are thus unable to multiply on their own and can only do so within a bacterium. They have a **RNA** or **DNA genome** they inject into the bacterium to replicate it using the cellular machinery. This injection is done through their tail, like a syringe. The bacteriophages, visualized in these photographs, at low and high magnifications, show an **icosahedral** head measuring about 1/10000 th of a millimeter and a tail 1/6000 th of a millimeter long.

Bacterio-phages

Bacteriophages are certainly the most numerous organisms on earth, on the order of 10,000 billion billion billion viral particles (10 ^{31}), and thus well beyond the number of bacteria. They are present everywhere, especially in seawater, where their concentration is estimated at ten billion per liter in surface seewater.

They are currently experiencing a resurgence of interest in the fight against bacteria multi-resistant to antibiotics, according to a principle called "phage therapy" or "phagotherapy". The use of bacteriophages specific to certain pathogenic bacteria could thus help to eliminate these pathogenic bacteria by killing them.

© The University of Tours 2025
P. Roingeard, *Journey to the Viral World: Electron Micrographs of Viruses*,
https://doi.org/10.1007/978-3-031-77995-4_14

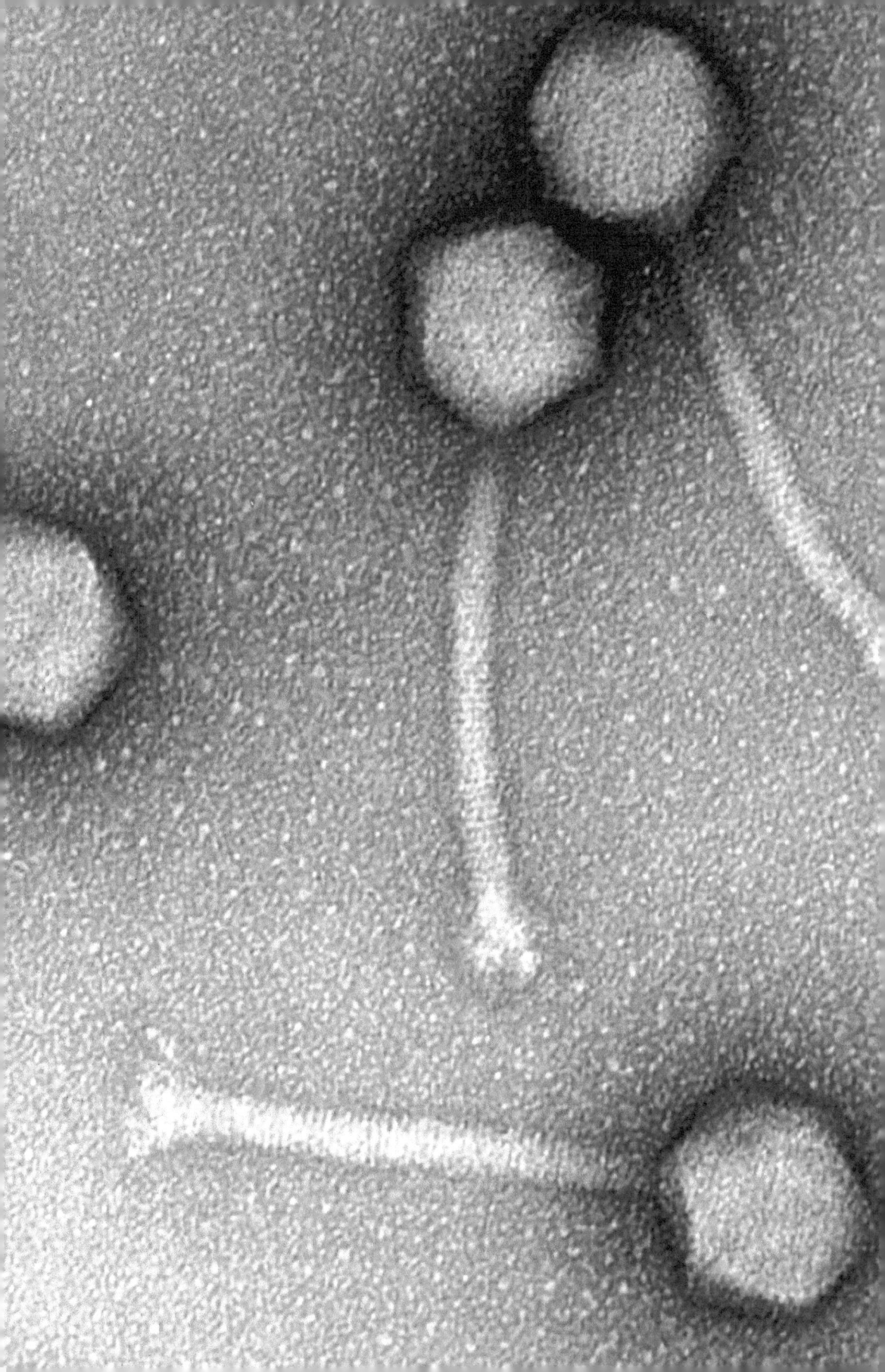

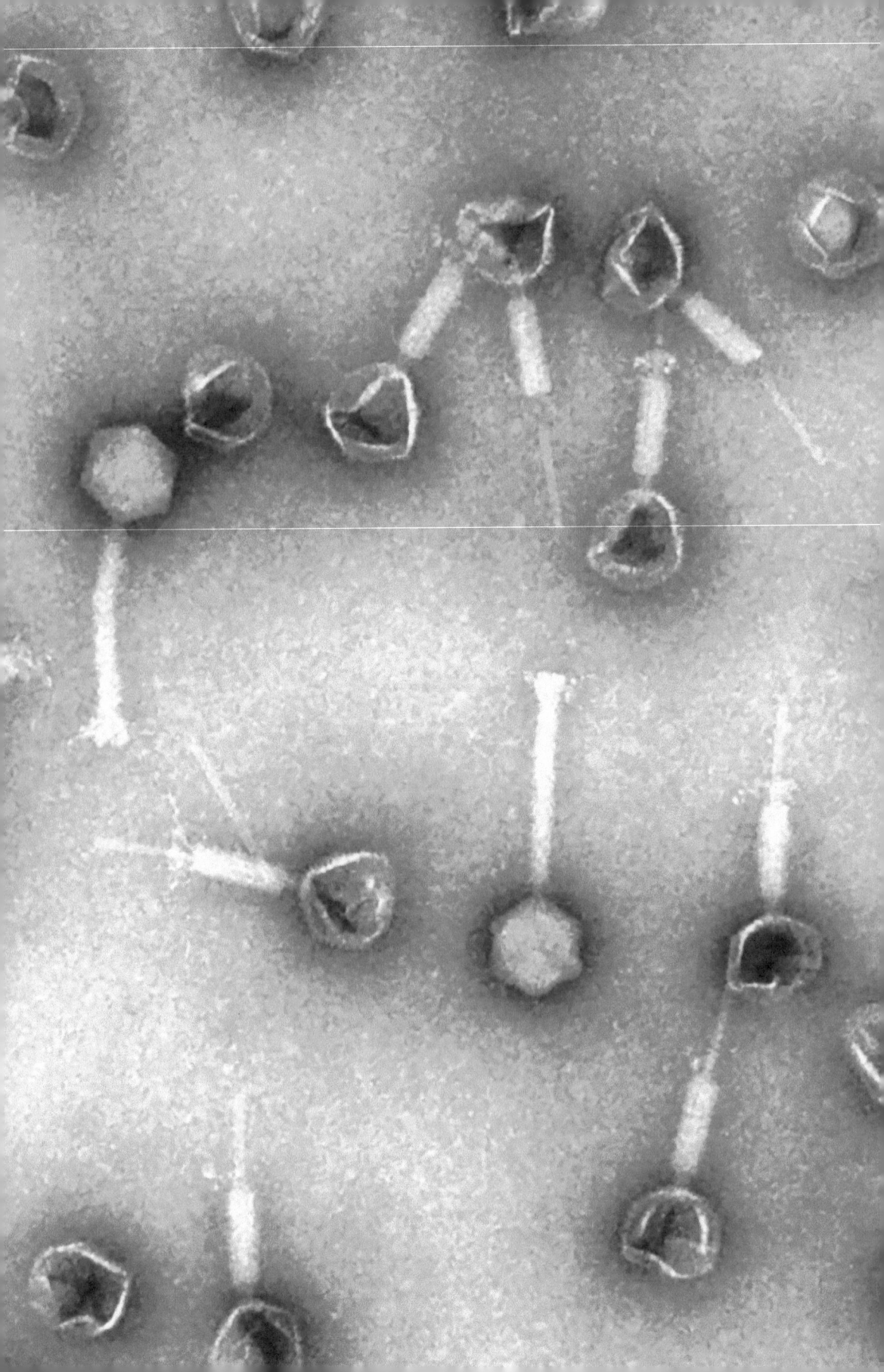

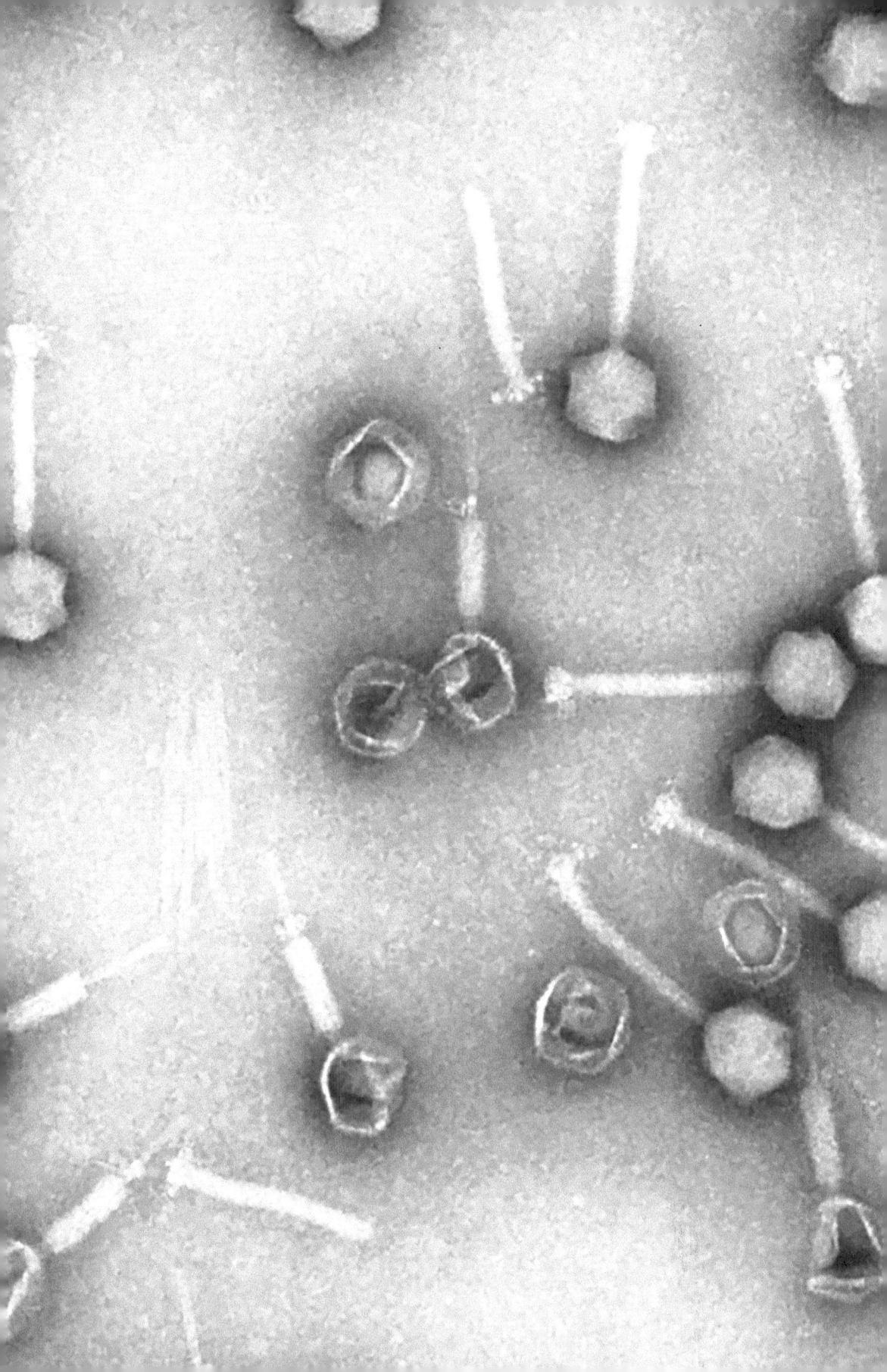

Panel 17. This image of a cell section illustrates the budding of a **helically** symmetrical enveloped virus on the surface of an infected cell. The term budding refers to the formation of a bud that spontaneously arises due to the grouping and assembly of viral proteins at the cell membrane. The viral bud forms at the cell membrane and individualizes to be finally released from the cell. During this mechanism, the viral **capsid** will be surrounded by a portion of the cell's plasma membrane, which then becomes the viral envelope.

The virus presented here is the Respiratory Syncytial Virus (RSV), which, as its name suggests, is responsible for respiratory infections. "**Syncytial**" refers to the fact that the virus is able to induce a *syncytium*, i.e. a fusion of infected cells (also see panel 33). The size of this virus varies from 1/8000th to 1/5000th of a millimeter. RSV infections are benign in adults, but can be serious in infants where they cause bronchiolitis. These are seasonal infections: in winter in northern countries, during the rainy season in southern countries.

Respiratory syncitial
virus budding

The colorized images bellow represent high magnifications of the RSV budding (left), and a virus that has completely detached from the cell membrane (right). The virus buds at the surface of the cell (in green), and it borrows a portion of the cell membrane (part in yellow). The **capsid** (in blue) is incorporated inside the viral particle. On such pictures, it is not possible to visualize the **helical** symmetry of the **capsid,** as only a tiny part of the capsid in present in the cell section. The "hairy" appearance of the virus is caused by the viral envelope proteins that are embedded in the portion of the membrane borrowed from the cell.

After viral budding, the cell membrane closes. Unless this phenomenon is too significant (which is rarely the case), this mechanism does not burst the cell. This is therefore a fundamental difference between naked viruses and enveloped viruses: naked viruses can only leave a cell by bursting it, while enveloped viruses do not necessarily burst the cell to be released. This does not mean that they are not harmful, but by other mechanisms. In the case of RSV, the pathology is mainly related to the immune response of the infected individual who will damage the infected cells.

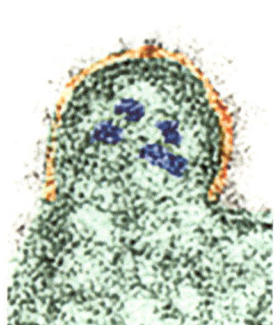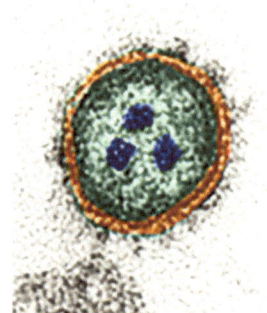

© The University of Tours 2025
P. Roingeard, *Journey to the Viral World: Electron Micrographs of Viruses,*
https://doi.org/10.1007/978-3-031-77995-4_15

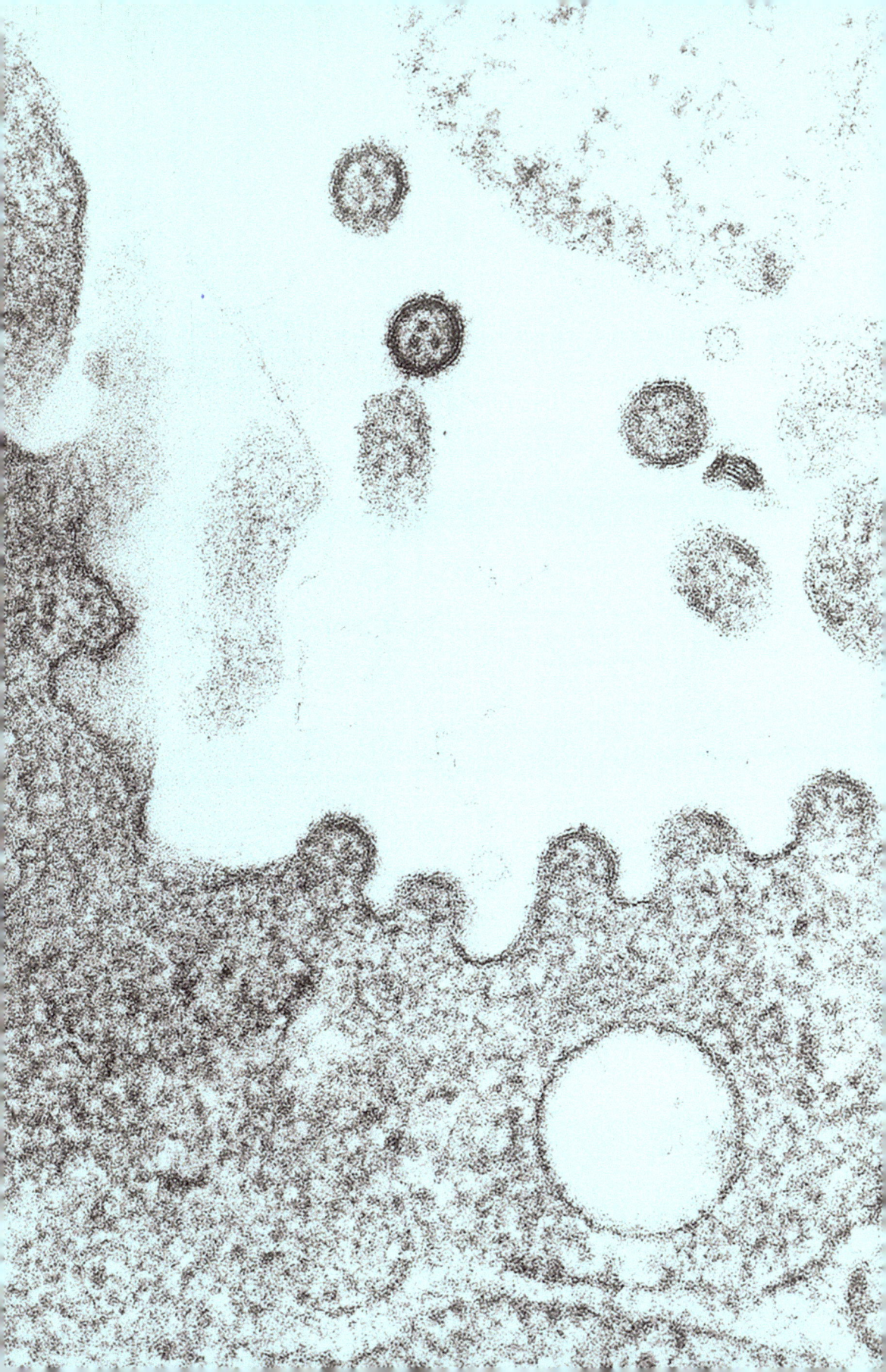

Panel 18. Not all enveloped viruses bud at the surface of the cell, borrowing a portion of the plasma membrane that delimits the cell and the extracellular environment. Some bud in compartments internal to the cell, also delimited by membranes, then exploit the cell's secretion mechanisms to exit the cell.

The herpes simplex virus (HSV) is particularly remarkable because it "travels" in several intracellular compartments. In the image at the bottom, 4 compartments of the cell, delimited by membranes, can be distinguished: the nucleus, at the very bottom; the perinuclear space (in green), delimited by the two concentric membranes surrounding the nucleus; cytoplasmic vesicles (in purple). About fifty viral **capsids** can be observed in

Herpes virus
the art of traveling
within the cell

the nucleus of this cell. One of them has just budded by surrounding itself with a portion of the first membrane encircling the nucleus, forming an enveloped particle present in the perinuclear space, in green. Another viral particle illustrates the next step, being present in a vesicle internal to the cytoplasm, in purple. A virus present in the perinuclear space can indeed loose its envelope, by fusion of this envelope with the second membrane encircling the nucleus, releasing its " de-enveloped " **capsid** into the cytoplasm. This **capsid** can then bud in an internal vesicle to surround itself again with an envelope made by the membrane of this vesicle (the situation of the virus in the vesicle colored in purple). These mechanisms of de-envelopment/re-envelopment are specific to herpes viruses. They are quite unique in the world of viruses.

The image at the top represents another area of this same cell and shows viruses that are in two distinct vesicles of the cytoplasm (in purple). On top left, viruses that have been released from the cell and now present in the extracellular environment (in light blue). To do this, the purple vesicles fuse with the plasma membrane of the cell to release their content into this extracellular environment. The HSV is a fairly large virus, on average 1/8000 th of a millimeter.

© The University of Tours 2025
P. Roingeard, *Journey to the Viral World: Electron Micrographs of Viruses*,
https://doi.org/10.1007/978-3-031-77995-4_16

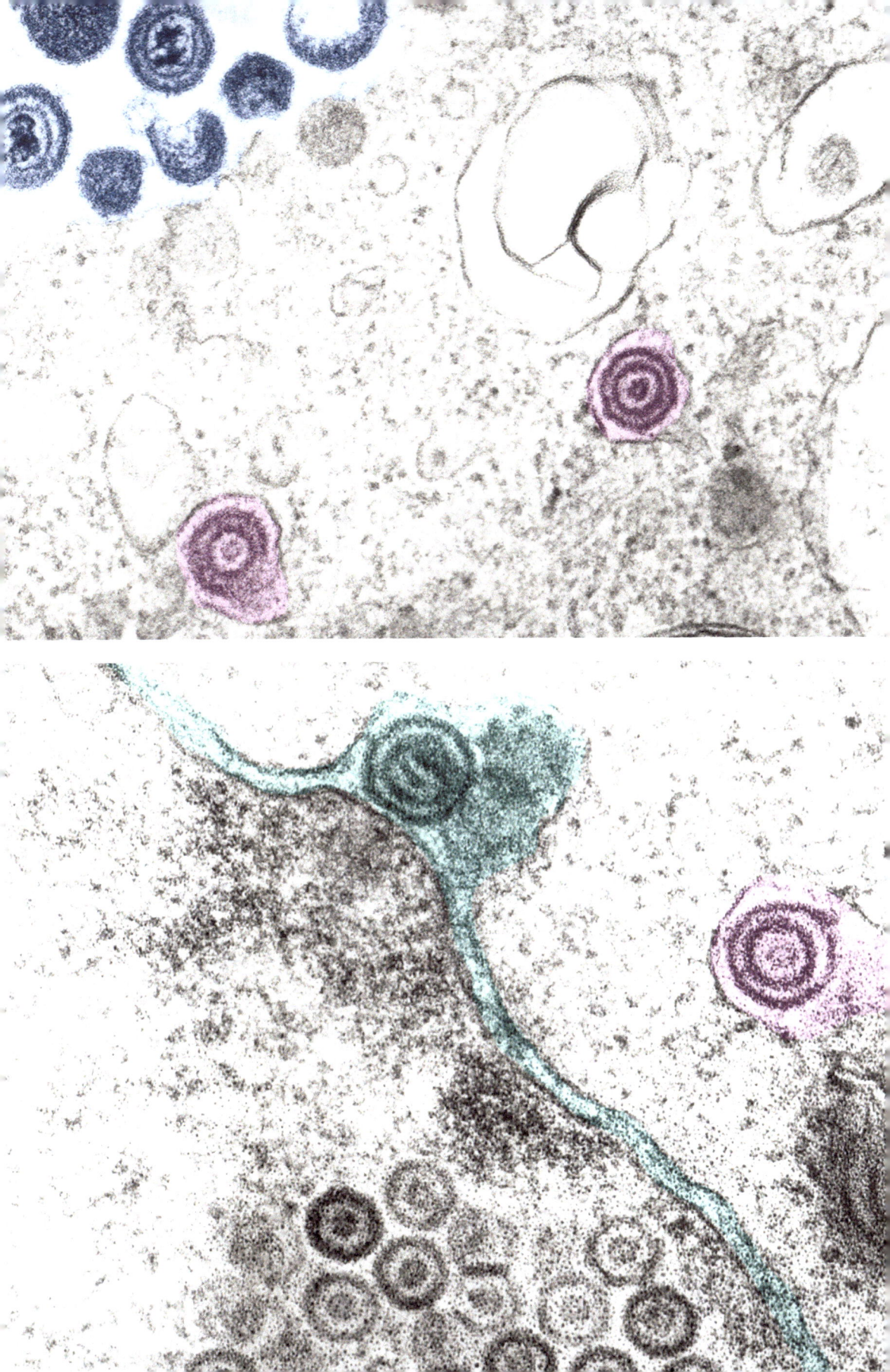

Panel 19. The herpes simplex virus (HSV), as well as other viruses (including the AIDS virus), are capable of establishing connections between two cells for rapid transmission of viral particles from one cell to another. This is referred to as a "viral synapse", similar to the synapse that represents a functional contact zone established between two neurons.

This image shows a close contact between two cells and an intense traffic of viruses between these two cells. The cell in purple is the "donor" cell: many viruses have just left it and are massively infecting the "recipient" cell, in green. This viral synapse allows for a very efficient infection because many viruses infect the cell at the same time and spend a minimum amount of time in the extracellular environment, thus escaping the immune system.

A viral synapse

There are two types of herpes simplex virus, HSV1 and HSV2. They are responsible for oral infections (cold sores) and sexually transmitted infections (genital herpes). They induce ulcerating vesicular eruptions on the skin and mucous membranes. These conditions are relatively benign in most subjects, but they can be severe in individuals with a weakened immune system, in newborns or in pregnant women, for whom the virus can spread throughout the body. This can lead to **encephalopathy**, which can be fatal if not treated in time.

A subject who has been infected with an HSV virus is never completely cured, as the virus can remain present in the body's cells in a latent phase, without inducing clinical signs. Reactivation episodes are possible following stress or episodes of **immunosuppression**. This decrease in immune system control then provides a favorable environment for the multiplication of the virus and the appearance of a new episode of vesicular eruptions. This phenomenon explains the name given to this virus, the word "herpes" coming from the ancient Greek *herpeton* (which crawls, reptile). Like the snake that sleeps or crawls and can awaken, the virus in the latent phase is capable of reactivating its infectious cycle.

© The University of Tours 2025
P. Roingeard, *Journey to the Viral World: Electron Micrographs of Viruses*,
https://doi.org/10.1007/978-3-031-77995-4_17

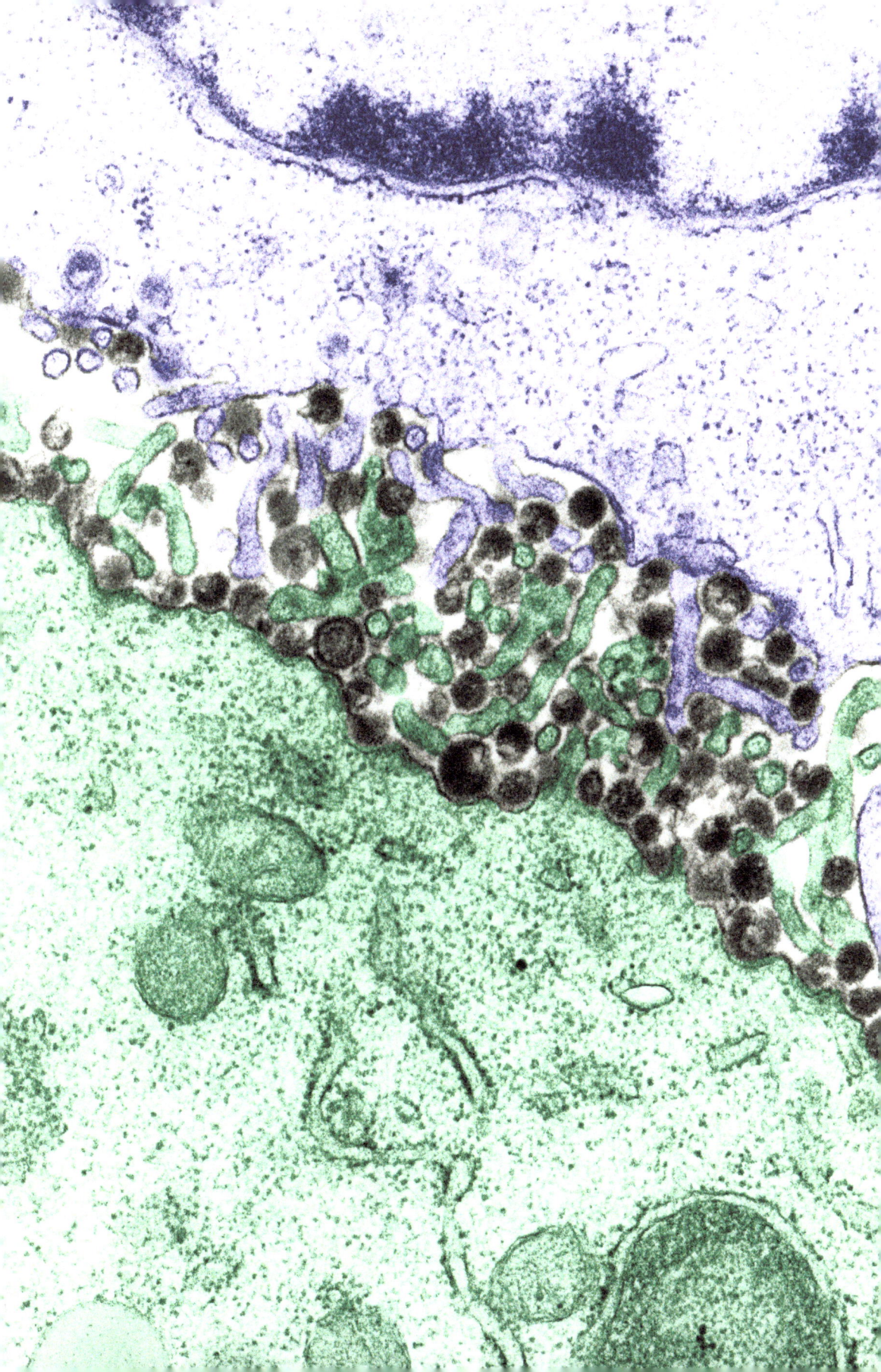

Panel 20. The Epstein-Barr virus (EBV) is part of the herpesvirus family. It is named after its two discoverers, Michael Epstein and Yvonne Barr, in the early 1960s. The image on left shows a cross-section of a cell massively infected by this virus: the cell is indeed literally covered on its surface by the viruses that have just being released. The image on right allows to visualize EBV at higher magnification, clearly showing that this is an enveloped virus containing a spherical **capsid**. However, unlike HSV and although part of the same family, the viral particle is much more compact and dark. It measures about 1/8000 th of a millimeter.

Nearly 90% of adults worldwide have been infected by EBV. Most of these infections have been mild and have gone unnoticed. They can occur during childhood, when the virus is transmitted by contact, or during adolescence when the virus can be transmitted between young adults through saliva. Following this infection, some of them develop a disease called infectious mononucleosis, often referred to as the "kissing disease". Mononucleosis results in a proliferation of lymphocytes, mononuclear blood cells (a cell with a single nucleus, unlike other blood cells). This explains the name given to this disease. It is character-ized by extreme fatigue, but

its course is most of the time favorable. However, this virus is associated with cancers in certain parts of the world, in conjunction with genetic and environmental factors. It can thus be responsible for lymphomas (cancerous proliferation of lymphocytes) and cancers of the upper aerodigestive tract.

Herpes viruses have been identified in many species outside of humans, including amphibians, fish, and even mollusks. There is indeed a herpes virus that is responsible for devastating epidemics in oyster farms. All these herpes viruses have a strict species barrier, which means that there is no risk for humans to consume infected oysters. However, this suggests that an ancestral herpes virus could have infected the animal kingdom well before its diversification into different branches. The herpes viruses would then have co-evolved with their host species.

© The University of Tours 2025
P. Roingeard, *Journey to the Viral World: Electron Micrographs of Viruses,*
https://doi.org/10.1007/978-3-031-77995-4_18

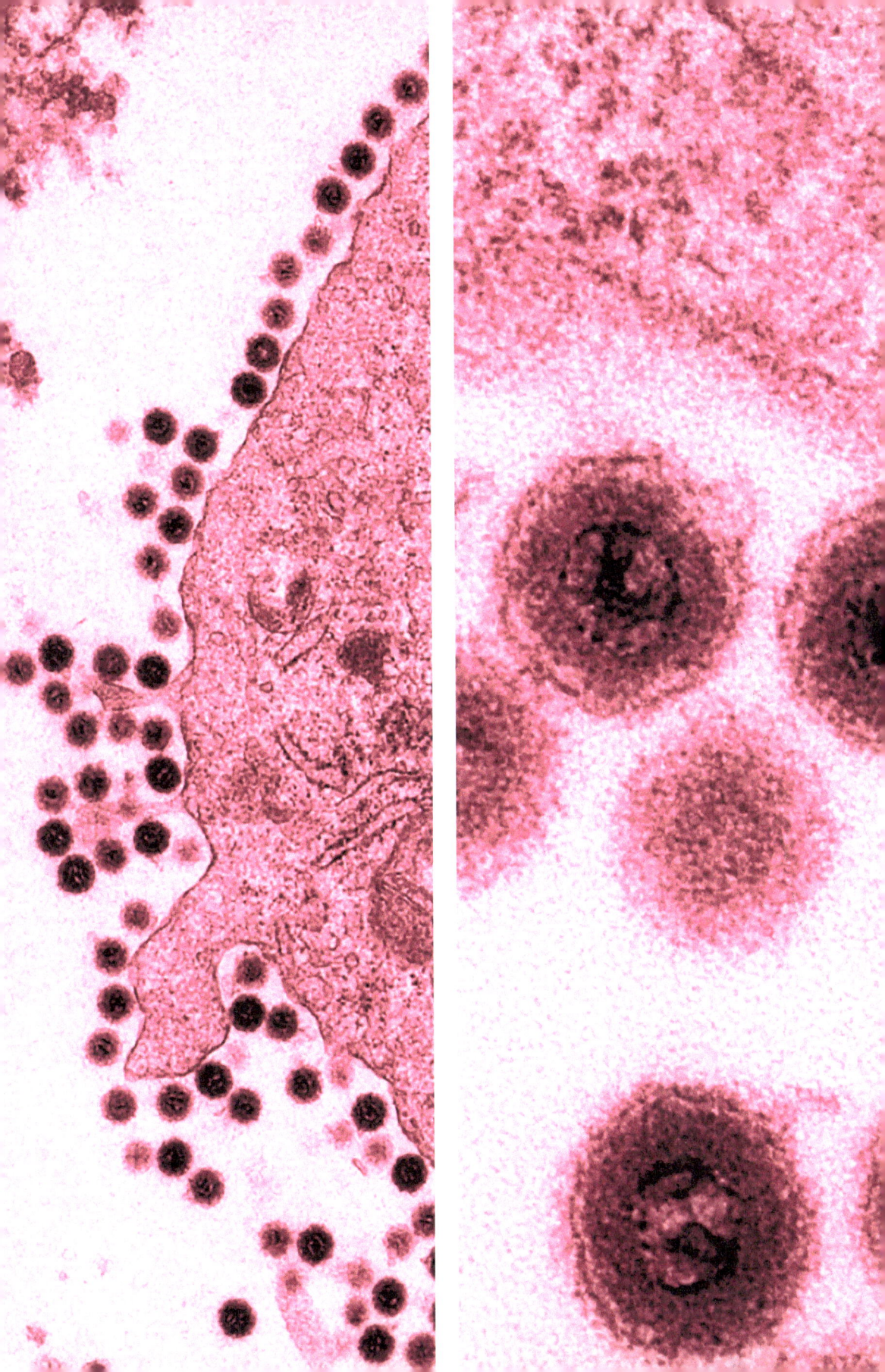

Panel 21. The entry of a virus into a cell always occurs through a specific receptor, similar to a key that corresponds to a very particular lock. The **capsid** proteins recognize the receptor on the target cell surface if the virus is naked, while the envelope proteins are responsible of such association in the case of an enveloped virus. This complementarity between a given viral protein and a receptor on the cell surface determines the virus's ability to infect a particular cell in the body. Thanks to this complementarity, the virus attaches to the surface of a cell, as here in the case of a herpes simplex virus (HSV).

Cells do not set up receptors on their surface intended to let in viruses, especially if they are pathogens. This would

Herpes Simplex Virus Entry

obviously be counterproductive. But the viruses exploit cellular receptors intended for important functions for the cell and hijack them for their own benefit. In the case of HSV, the virus attaches to the cell surface on a receptor that normally serves for adhesion between cells to maintain the architecture of tissues. This image shows the close association between the virus envelope and the cell membrane in this process of recognizing a surface receptor.

The next step for the virus is to fuse its envelope with the plasma membrane of the cell, in order to release its **capsid** into the cytoplasm of the cell. The virus then introduces its **genome** into the cell, and can then replicate.

© The University of Tours 2025
P. Roingeard, *Journey to the Viral World: Electron Micrographs of Viruses*
https://doi.org/10.1007/978-3-031-77995-4_19

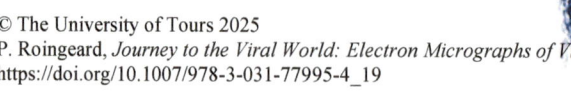

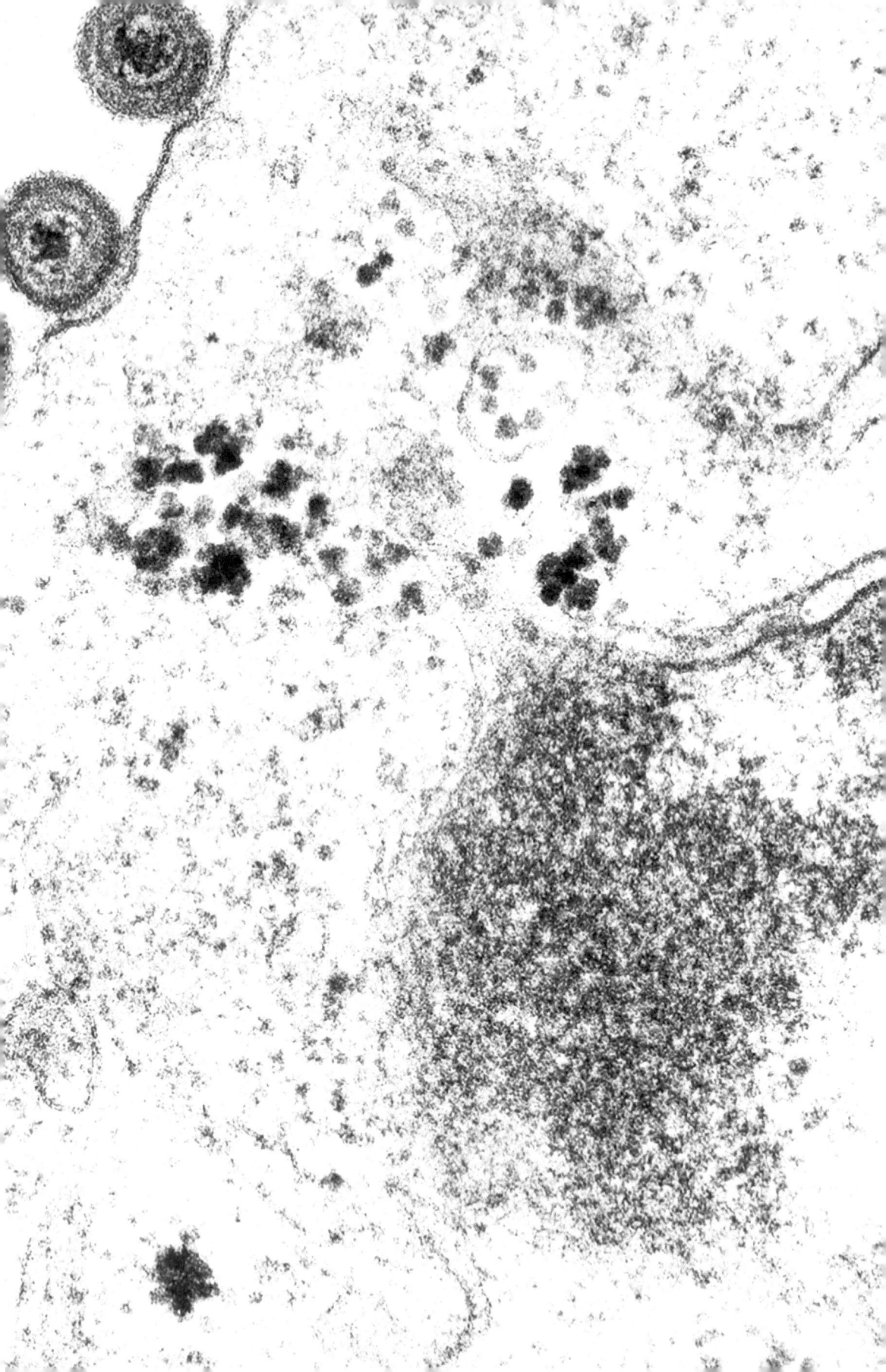

Panel 22. The varicella-zoster virus (VZV) is also part of the herpes virus family. It is one of the most contagious viruses for humans. It is easily transmitted through the respiratory route, via droplets of secretions projected (coughing, sneezing…) from an infected individual.

The disease, also called "chickenpox"(also see panel 44) is one of the most common in early childhood and is characterized by vesicular eruptions, which are fortunately benign. After an episode of infection, a vigorous immune response prevents any reinfection. However, as with other herpes viruses, an individual who has been infected with the VZV never completely eliminates the virus, which can remain silent for years. Episodes of reactivation of this virus are rare, but will result in shingles, characterized by skin eruptions causing intense burning sensations.

There is a vaccine against this virus, but it is not systematically used to vaccinate children. Indeed, in the case of this virus which is not very pathogenic for children, it is preferable to let the virus circulate to induce good immunity within the population. Insufficient vaccine coverage could risk shifting the onset of this disease to adulthood and then lead to much more severe forms. The vaccine is thus reserved for teenagers and pregnant women who have not encountered the virus.

Varicella-Zoster Virus Entry

This photograph shows a VZV entering a cell. An image with this level of detail is quite rare as it shows all the components of the virus: an **icosahedral capsid**, for which the angles of the **icosahedron** are visible; the **DNA genome** of the virus within this **capsid**; a lipid envelope surrounding the **capsid**, in which the envelope proteins of the virus are embedded (giving a "heary" appearance to the surface of the viral particle). The interaction of the viral envelope proteins with the receptors present at the cell membrane level can be also visualized. The next step will be the fusion of the virus envelope with the cell membrane, thus releasing the **capsid** of the virus into the cytoplasm of the cell.

© The University of Tours 2025
P. Roingeard, *Journey to the Viral World: Electron Micrographs of Viruses*,
https://doi.org/10.1007/978-3-031-77995-4_20

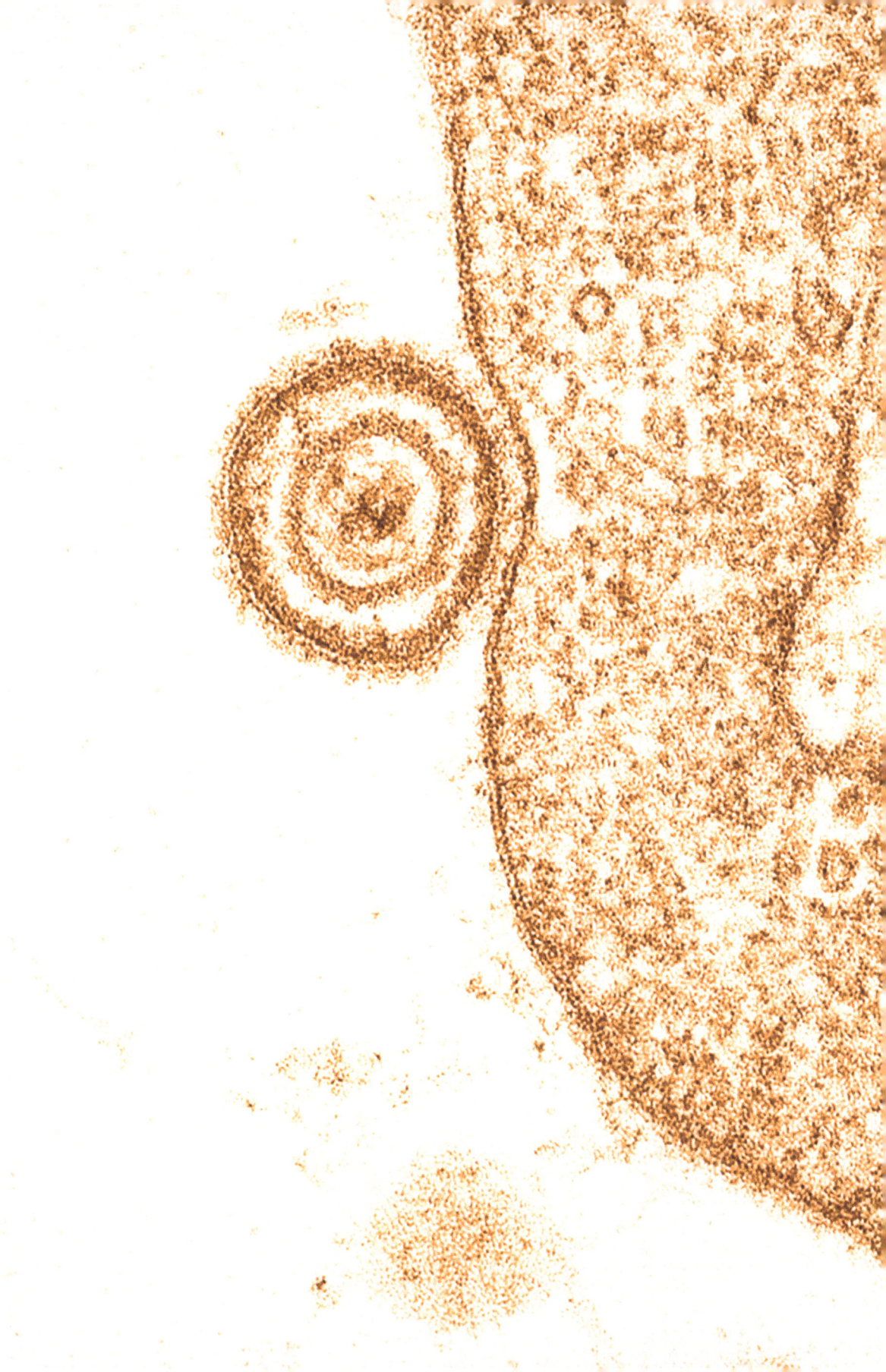

Panel 23. Many viruses, like the dengue virus (also see panel 54), enter the cell by endocytosis. Endocytosis is a physiological mechanism to allow molecules present in the extracellular environment to penetrate within the cells. The cell (in green) curves its membrane towards its cytoplasm to form a vesicle that will individualize. The cell thus captures extracellular fluid that contains nutrients, but also certain molecules that can specifically enter the cell via a receptor. For example, the cell allows by such mechanism the entry of transferrin, a protein that regulates the concentration of iron inside the cell.

Viruses can exploit this transport mechanism, very common on the surface of cells, to penetrate inside a cell. These three images show how the dengue virus enters a cell: at the left, the virus recognizes a receptor on the cell surface and the cell membrane begins to curve; in the middle, the membrane then deforms strongly to encircle the

Dengue virus entry

viral particle; at the right, the vesicle is closed and thus completely internalized within the cell. In the case of the dengue virus, which is a small enveloped virus, the virus envelope fuses with the membrane of this vesicle to release its **capsid** into the cytoplasm of the cell.

In the case of naked viruses, without an envelope, their strategy is to burst the endocytosis vesicle to allow their **capsid** to be released into the cellular cytoplasm. In the end, the consequences of these two different strategies are the same; the **capsid** of the virus ends up in the cytoplasm of the cell and the virus can initiate its cycle of multiplication.

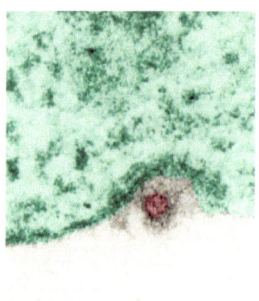 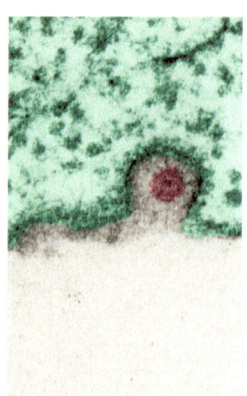 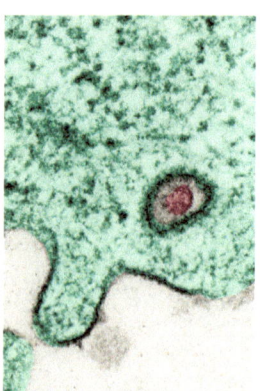

P. Roingeard, *Journey to the Viral World: Electron Micrographs of Viruses*, https://doi.org/10.1007/978-3-031-77995-4_21

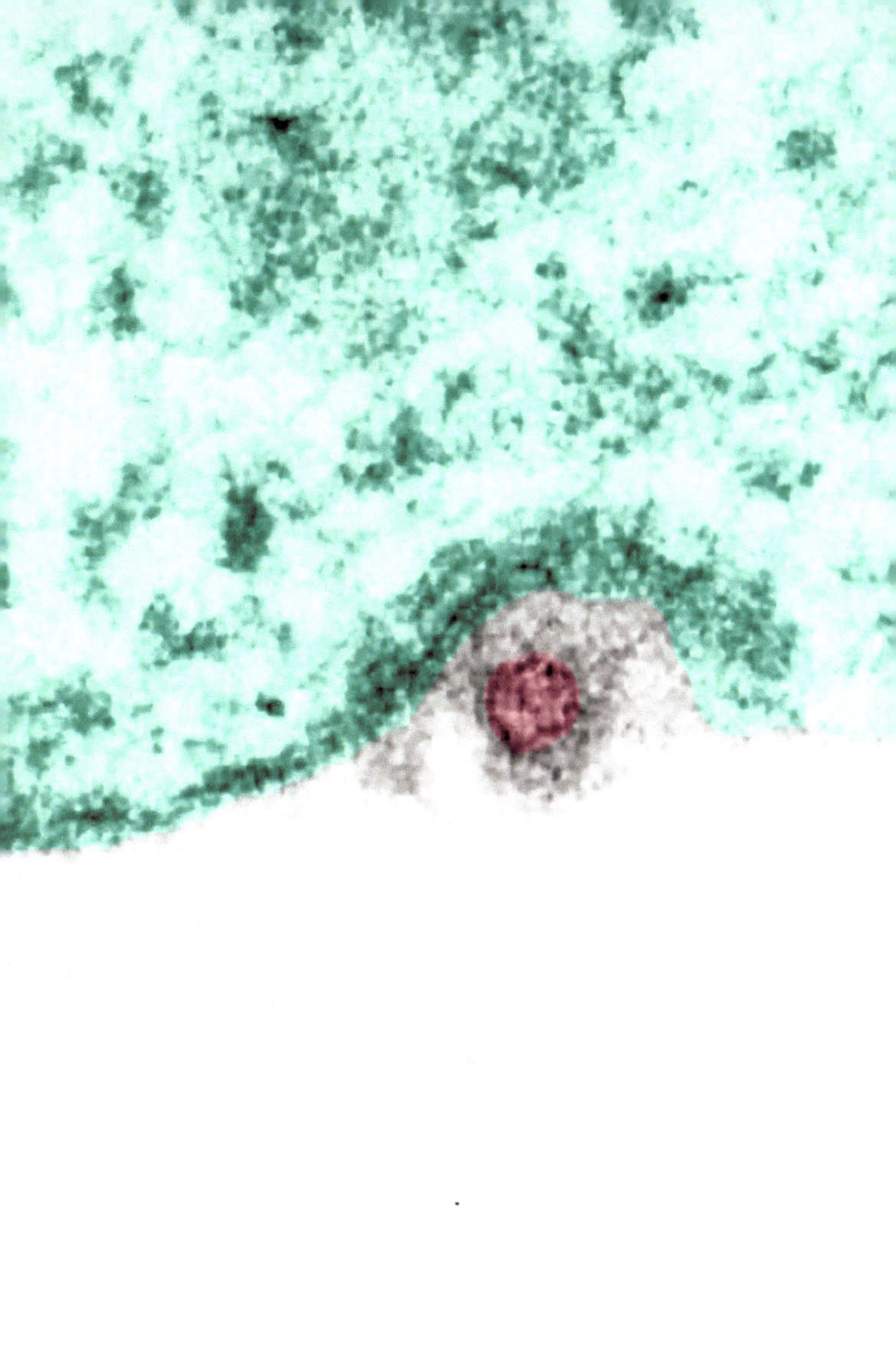

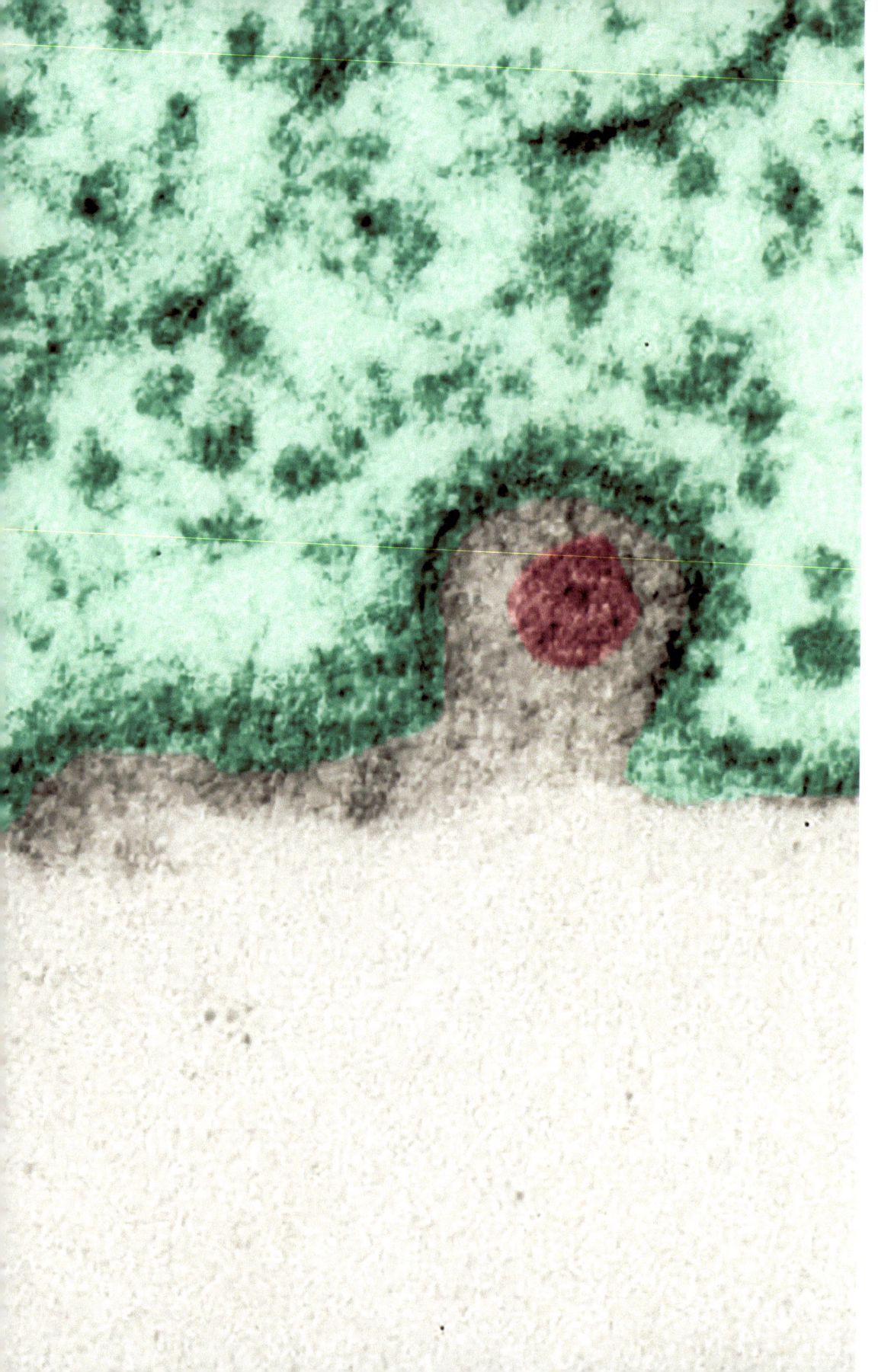

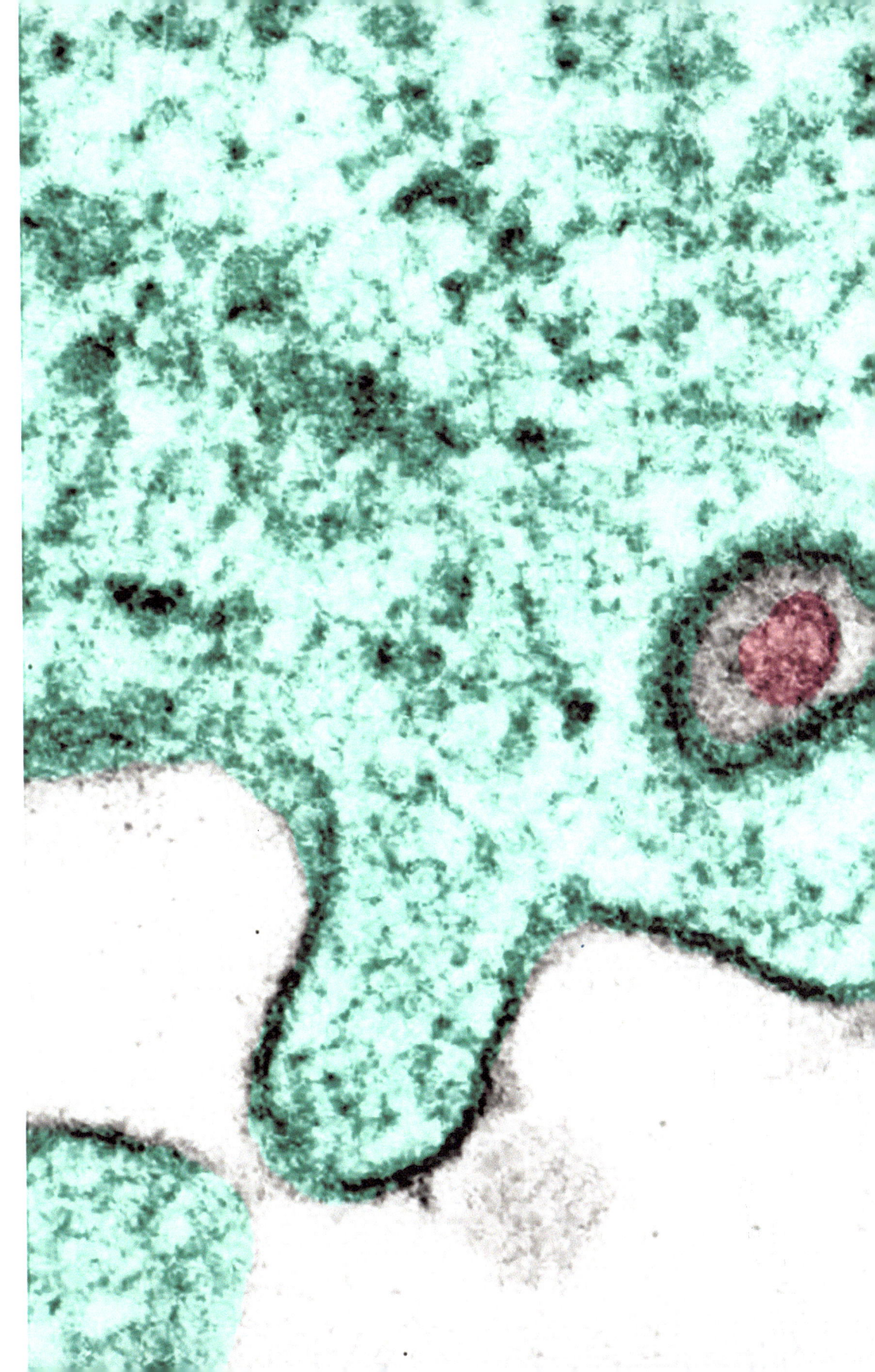

Panel 24. After a virus enters a cell, the **capsid** is released in the cytoplasm and can exploit the cell's constituents to move within the cell. These photographs show "foamy" viruses. The inset shows the complete virus, consisting of a **capsid** (in orange) surrounded by an envelope (in yellow). This virus enters the cell through endocytosis vesicles, and its envelope fuses with the membrane of these vesicles to release the viral **capsid** inside the cell. Once in the cell, this **capsid** moves within the cell using the cellular microtubules. Microtubules (as their name suggests, literally "microscopic tubes") are very important structures in

a cell. They constitute a sort of highway for the circulation of cellular components. Organized like the spokes of a bicycle wheel, theses tubes radiate from a structure called the cell center, formed by two cylinders themselves made up of numerous embedded microtubules.

Intracellular traffic of a viral capsid
the foamy virus

The colorized image opposite shows a foamy virus **capsid** (in orange) moving along these microtubules (in green). This **capsid** is surrounded by cellular proteins (gray structures forming spikes around the **capsid**) that help it slide along the microtubules. Thanks to these mechanisms, the viral **capsids** are adressed rapidly around the cell center, bringing them closer to the cell nucleus. The image in black and white, at lower magnification, shows a large number of viral **capsids** grouped around the cell center, consisting of two cylinders (one in cross-section, the other in longitudinal section). This strategy, which consists of exploiting cellular communication mechanisms, allows the capsids to arrive more quickly around the nucleus, into which they must enter to initiate the viral **genome** replication. This constitutes another illustration of the hijacking of cellular mechanisms by a virus for its own benefit.

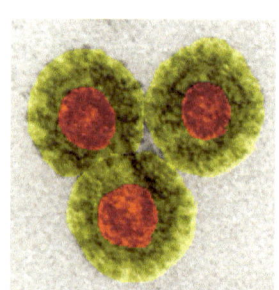

P. Roingeard, *Journey to the Viral World: Electron Micrographs of Viruses*, https://doi.org/10.1007/978-3-031-77995-4_22

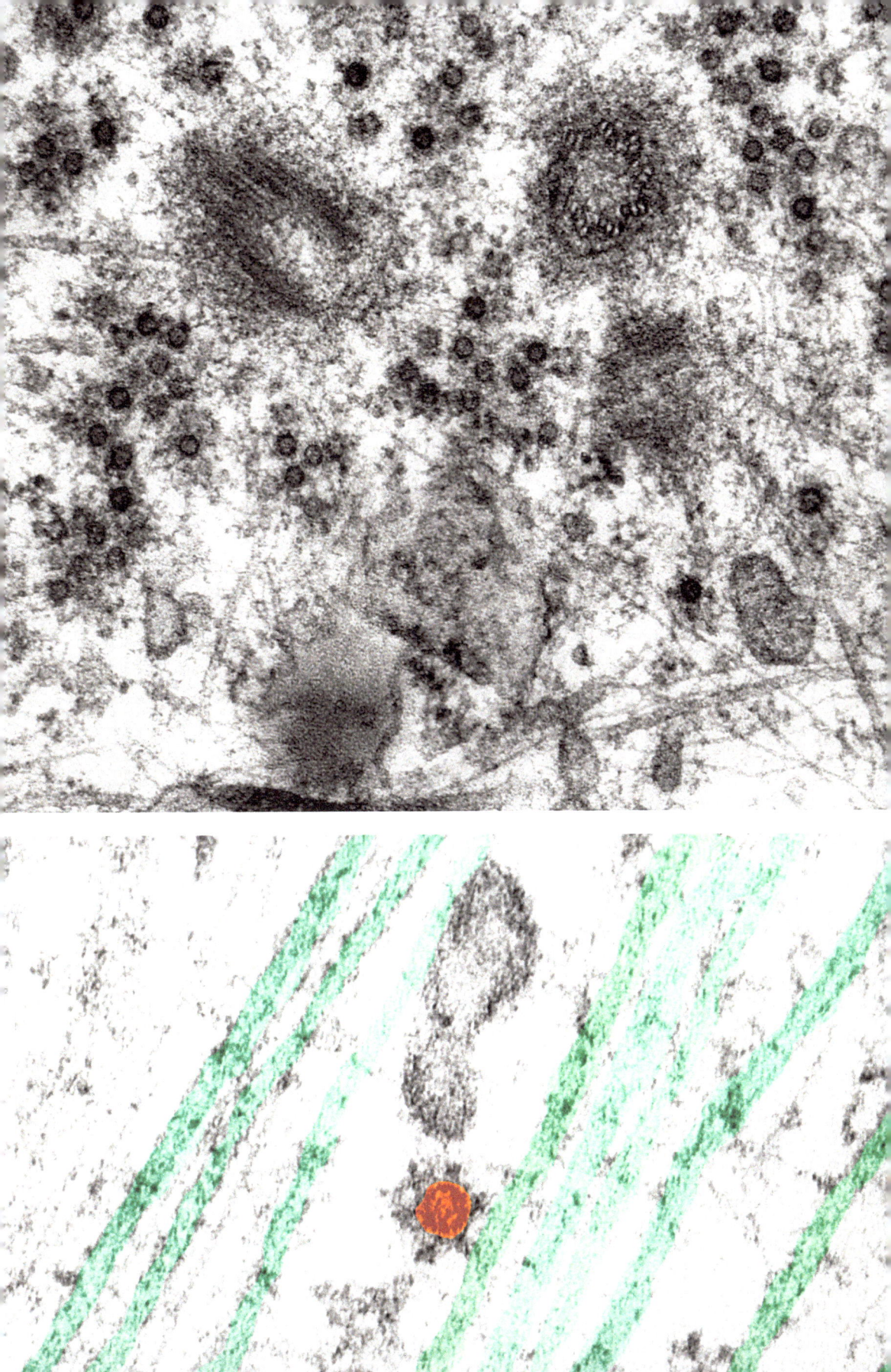

Panel 25. The name of the foamy virus comes from the fact that this virus induces sort of "bubbles" in infected cells, giving them a foamy appearance. This virus has been known since the 1950s, when it was first identified in non-human primates. The question of the existence of a foamy virus in humans has long been debated. In fact, it turns out that the few infections in humans by a foamy virus were the result of a fortuitous transmission from an animal to a human, with no possibility of subsequent human-to-human transmission. Foamy viruses have now been identified in felines and bovines, and the few cases

of transmission to humans have occurred in particularly exposed subjects (animal technicians, veterinarians, hunters…). No pathology related to these infections has been clearly identified in animals or humans. However, this virus represents an important study model for studying virus/cell interactions in the laboratory. In this photograph of an infected cell section, viral particles are seen forming by budding on the cell surface.

The foamy virus is part of the large family of **retroviruses**. This name of **retrovirus** comes from the fact that these viruses, which have an **RNA genome**, must perform a **reverse transcription** of their **genome** into **DNA** to set up their multiplication cycle. This **viral DNA** then integrates into the host cell's **DNA**, potentially disrupting the chromosomes. Between the 1950s and 1970s, numerous **retroviruses** were identified in animal species as varied as rodents, birds, and primates. At that time, it was thought that there were no **retroviruses** in humans, until the discovery of HTLV (for Human T Lymphotropic Virus) in 1981 (see panel 27), and then HIV (Human **Immunodeficiency** Virus) in 1983 (see panels 28 to 32). It is very likely that these two human **retroviruses** originated from a transmission of non-human primate viruses to humans, but unlike the foamy viruses, these viruses were able to adapt to humans and then spread widely in populations through human-to-human transmission. In the case of HIV, it has been estimated, through analyses of the **genomes** of non-human primate **retroviruses** and circulating human viruses, that this species barrier crossing probably occurred at the end of the 19th century.

© The University of Tours 2025
P. Roingeard, *Journey to the Viral World: Electron Micrographs of Viruses*,
https://doi.org/10.1007/978-3-031-77995-4_23

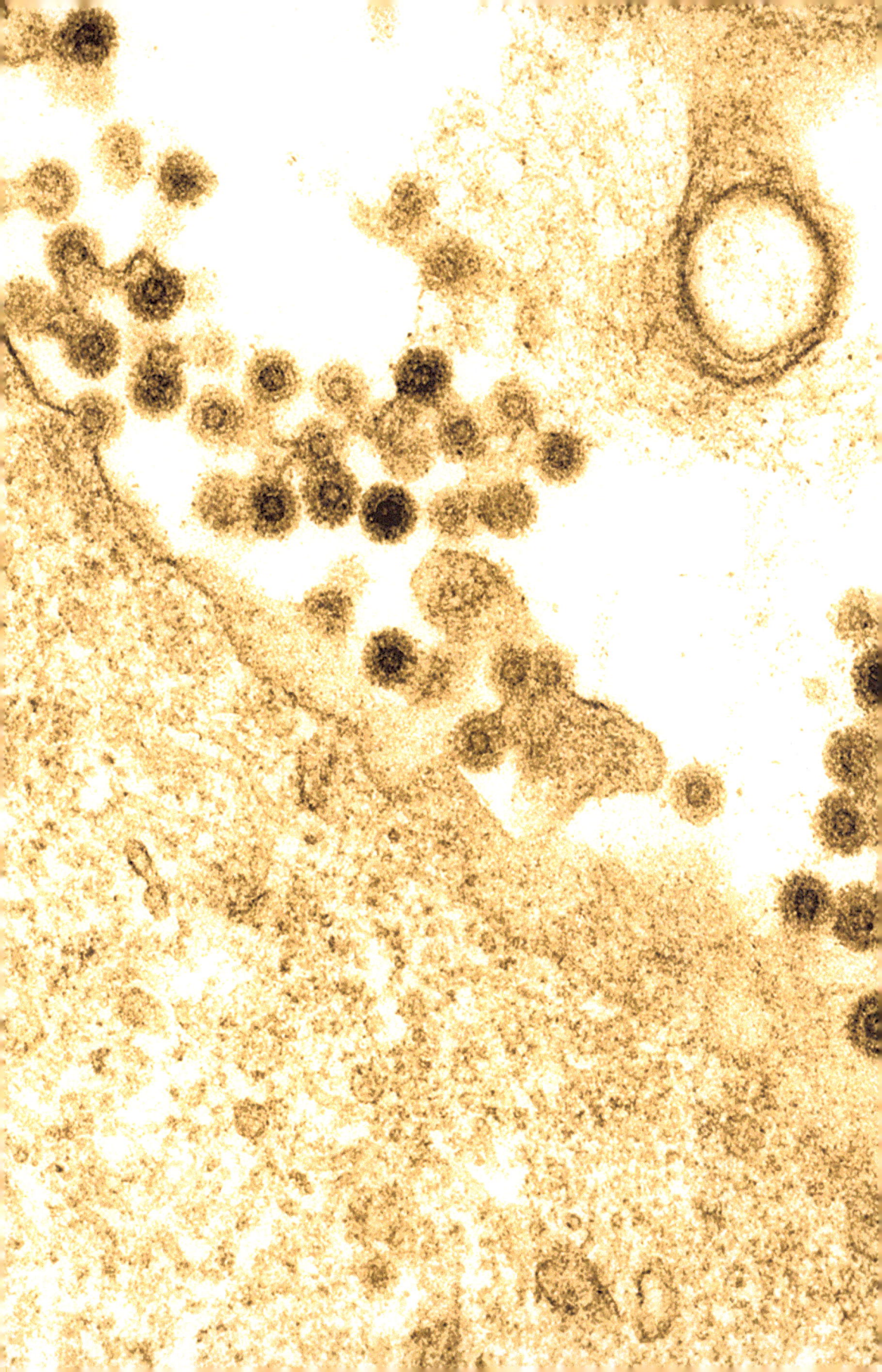

Panel 26. The MLV, for Murine Leukemia Virus, represents one of the many **retroviruses** identified in animals during the 1950s. These two images at two different magnifications show MLVs released by an infected cell. The inset image shows isolated MLVs in solution, observed by negative staining. The virus is about 1/10000 th of a millimeter in diameter.

As its name suggests ("murine" means related to mice), this virus is responsible for the development of leukemia in mice. Many of these **retroviruses** have thus been identified as associated with the development of cancers in the animals they infect. They have indeed constituted very important models for understanding the mechanisms of carcinogenesis, whether these cancers are caused by viruses or by other mechanisms such as exposure to chemicals or radiation.

An animal retrovirus
the MLV

Some animal **retroviruses** carry in their **genome** all the genetic information to make cells cancerous. They host an **oncogene** (literally, a gene that has the potential to cause cancer) which makes them capable of very quickly transforming the infected cell into a cancerous cell. We now know that these genes are of cellular origin. They were randomly captured through recombinations with the **genome** of the virus during evolution.

Other **retroviruses** transform normal cells into cancerous cells much more slowly. The transformation is then linked to the integration of the viral **DNA** into the chromosomes of the cell, which can cause a disruption of the cell's functioning and make it cancerous in the long term.

P. Roingeard, *Journey to the Viral World: Electron Micrographs of Viruses*,
https://doi.org/10.1007/978-3-031-77995-4_24

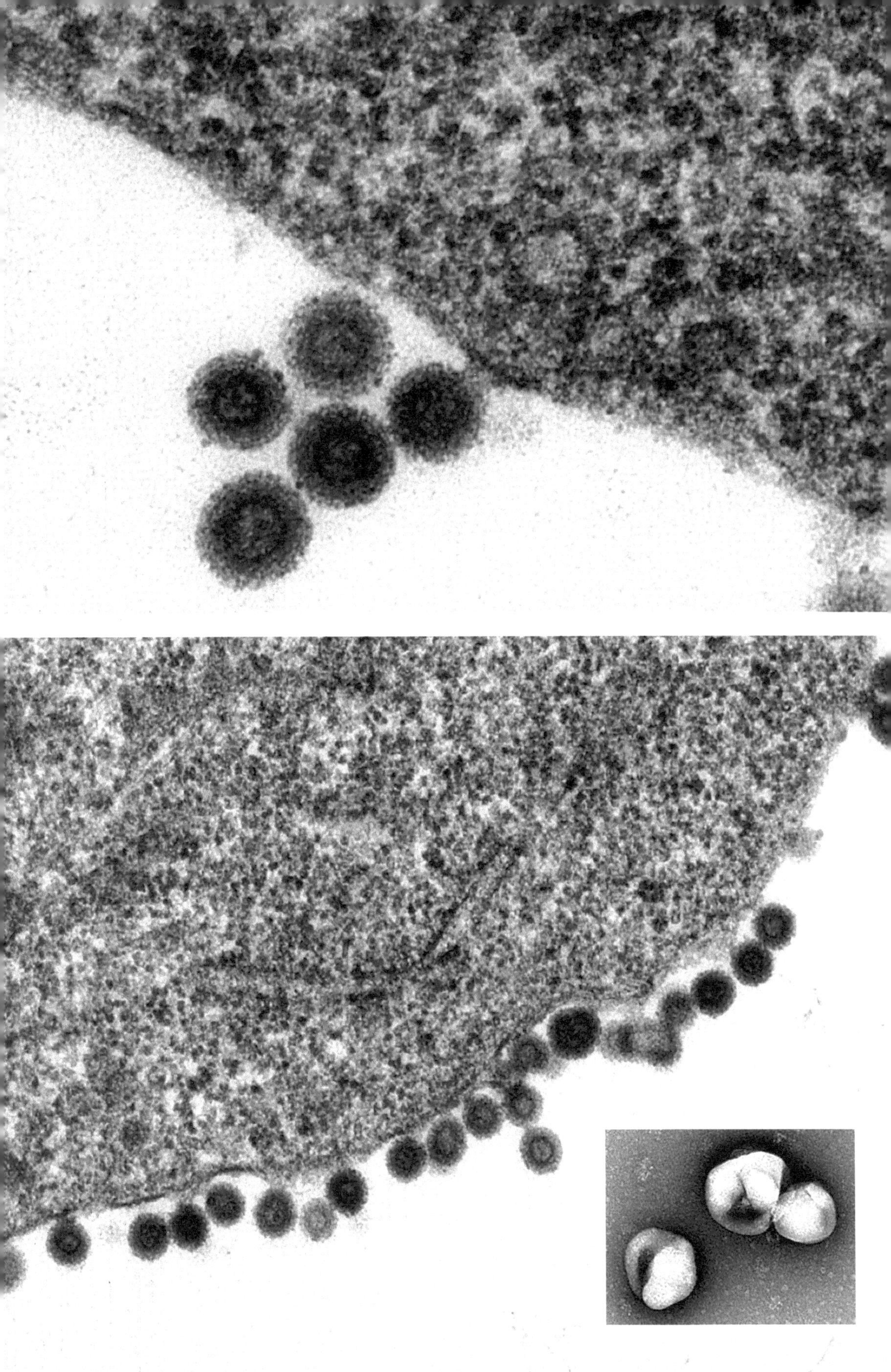

The HTLV

Panel 27. The HTLV (for Human T Lymphotropic Virus) represents the first **retrovirus** identified in humans, in 1981. With a size about 1/10000 th of a millimeter, it has a heterogeneous morphology. It is composed of a large **capsid** (in brown in the inset, surrounded by an envelope, in blue). This virus infects between 10 and 20 million people worldwide, especially in Japan, Central Africa, and the Caribbean. Chronic infection with this virus goes unnoticed, but it can be associated in the long term (10 to 20 years) in some individuals with leukemias or lymphomas. The causes of the development of these diseases are not well known, but they are likely the result of environmental factors combined with chronic infection by the virus. The virus does not have an **oncogene**, but the integration of its **genome** into cellular chromosomes can cause cancerous proliferations in the long term.

In the Caribbean, chronic infection with HTLV can also lead to a disease called "Tropical Spastic Paraparesis", characterized by lower back pain, lower limb pain, and gait disorders.

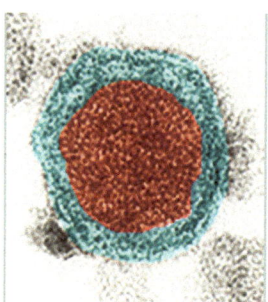

© The University of Tours 2025
P. Roingeard, *Journey to the Viral World: Electron Micrographs of Viruses,*
https://doi.org/10.1007/978-3-031-77995-4_25

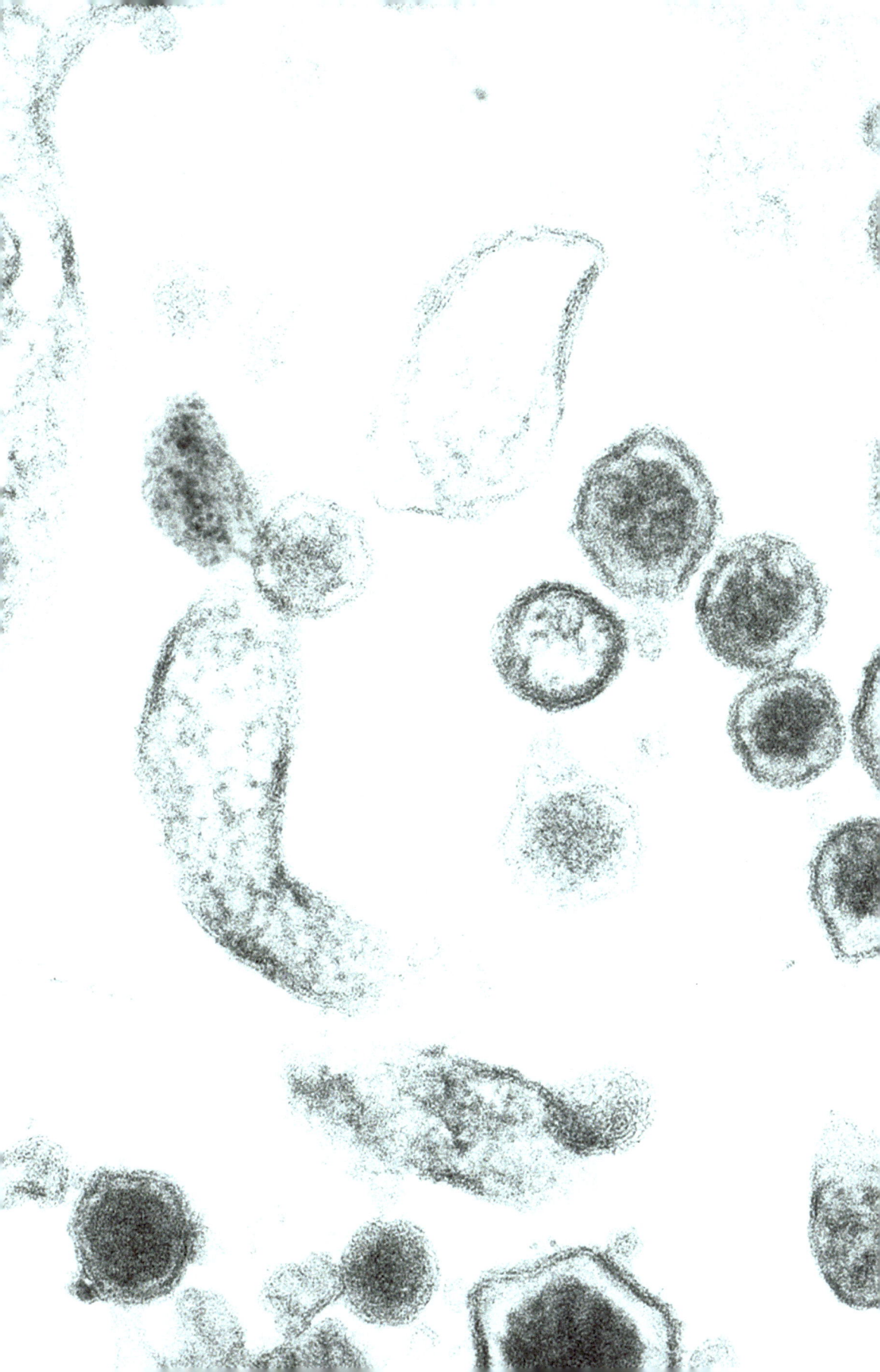

Panel 28. Shortly after the discovery of HTLV, a second human **retrovirus** was identified in 1983, within the community of men who have sex with men, in which this virus had rapidly spread. It was unknown at that time that the epidemic induced by this virus would have a significant impact on our society and constitute one of the major events of the late 20 th century. Since the

The AIDS virus

beginning of the epidemic, it was estimated in 2023 that over 84 million people have been infected with HIV, among which 40 million have died from the disease related to this infection. Sexual trans-

mission is not the only cause of transmission as the virus can also be transmitted through blood or breast milk.

While many animal **retroviruses** can make cells cancerous, HIV has the completely opposite effect, as it kills the infected cells. As the target cells of the virus are cells of the immune system, the CD4+ T lymphocytes, it is thus responsible for a pathology called AIDS (for Acquired **Immuno-Deficiency** Syndrome). If untreated, infected individuals develop numerous infections (so-called "opportunistic" infections) against which the body, deprived of its immune system, cannot fight.

These images show HIVs that have just been released from an infected cell. Unlike HTLV, the virus contains a much more compact **capsid** within the viral particle. Depending on the position of the virus in the section, this **capsid** can either take the form of a circular structure giving the virus the appearance of a target with a slightly off-center heart, or the form of a bar giving the virus the image of a "no entry" sign. This particularity of HIV is due to a specific mechanism of maturation of its viral **capsid** (decribed in the next panel).

© The University of Tours 2025
P. Roingeard, *Journey to the Viral World: Electron Micrographs of Viruses*,
https://doi.org/10.1007/978-3-031-77995-4_26

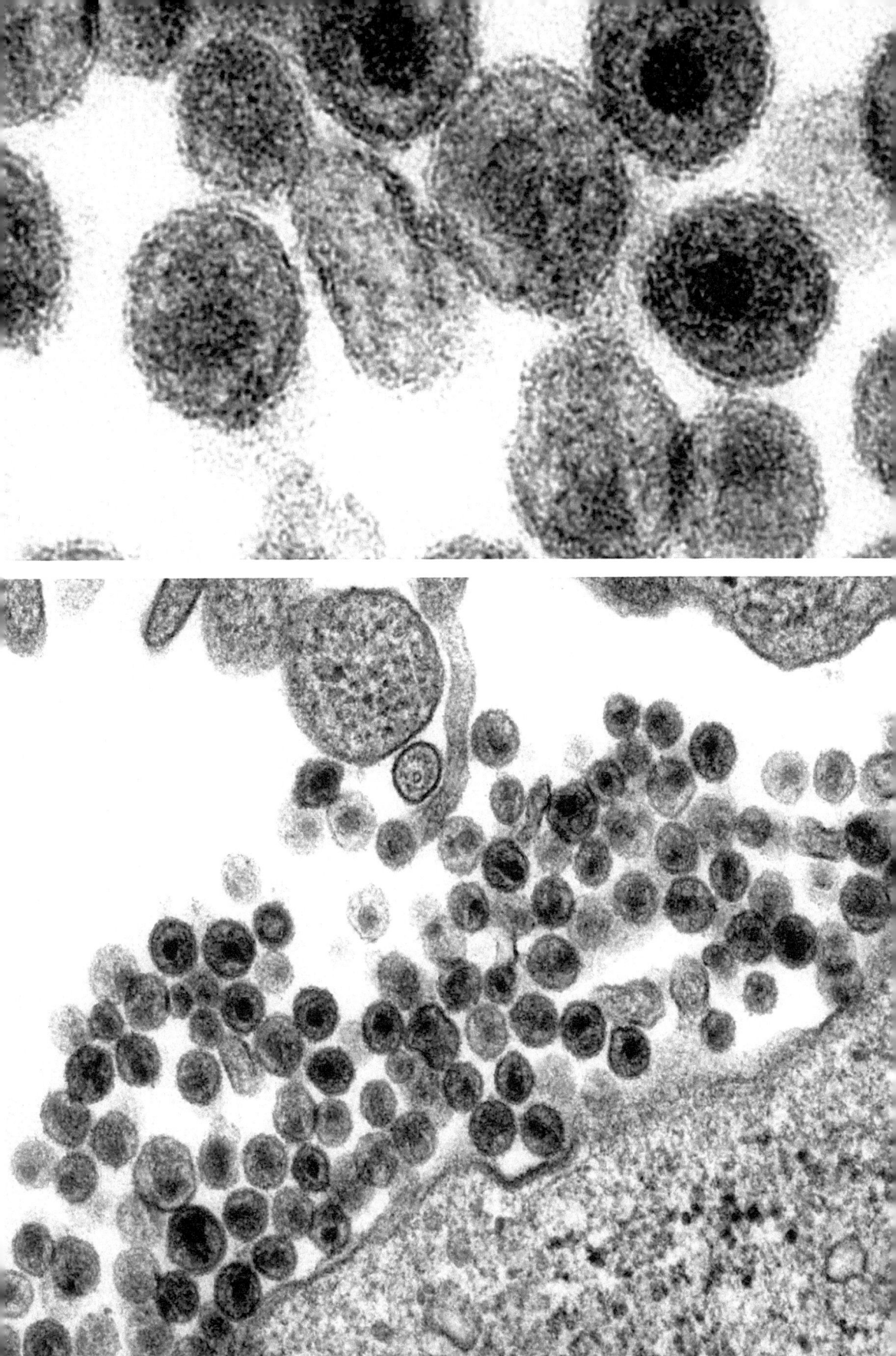

Panel 29. This image shows 4 HIVs in the process or just after exiting an infected cell (in green). It has the merit of showing, from top to bottom, the successive stages of the maturation of the viral **capsid**. In the first position, at the very top, the virus buds out of the cell thanks to its **capsid** protein (in red) which forms a sphere at the level of the cell's plasma membrane. This sphere extrudes from the cell, surrounding itself with a portion of the cell's plasma membrane (in yellow). At this stage, the virus is still in contact with the cell, it is just held back by a "neck". In the second position, a narrowing of this "neck" constitutes

a preliminary step to the rupture of this link between the viral particle and the cell. In the third position, this virus is completely detached, the cell's membrane having closed behind it. The **capsid** then forms a complete sphere, but in this case, the viral particle is not infectious. For the virus to become infectious and be able to infect a new cell, this **capsid** must undergo a maturation step. This is visualized on the viral particle in the fourth position, at the very bottom, where a compacted **capsid** with the apperarance of a "no entry" sign is observed. This reorganization of the **capsid** is induced by a viral enzyme that cuts the **capsid** protein (a protease) to give it this shape. Without the action of this protease, the virus is therefore not infectious. This particular mechanism is the target of certain antiviral drugs, the protease inhibitors, which block the action of this enzyme.

These protease inhibitors are not the only drugs that can be used to act against the virus. Others block an enzyme that is responsible for the integration of its **DNA** into the cell's chromosomes (integrase inhibitors). Some prevent the viral particle from entering the cell (inhibitors of the fusion of the viral envelope with the cell membrane). Unfortunately, access to care, and especially to these antiviral molecules, is not always optimal in some of the world's poorest countries. In 2023, it was estimated that about 40 million people worldwide were infected with this virus, many without access to these antivirals. Moreover, nearly 2 million new infections occur each year worldwide, including about 200,000 in newborns who get infected through their mothers' contaminated milk. Nearly a million people die from AIDS-related complications each year (2 people every minute).

© The University of Tours 2025
P. Roingeard, *Journey to the Viral World: Electron Micrographs of Viruses,*
https://doi.org/10.1007/978-3-031-77995-4_27

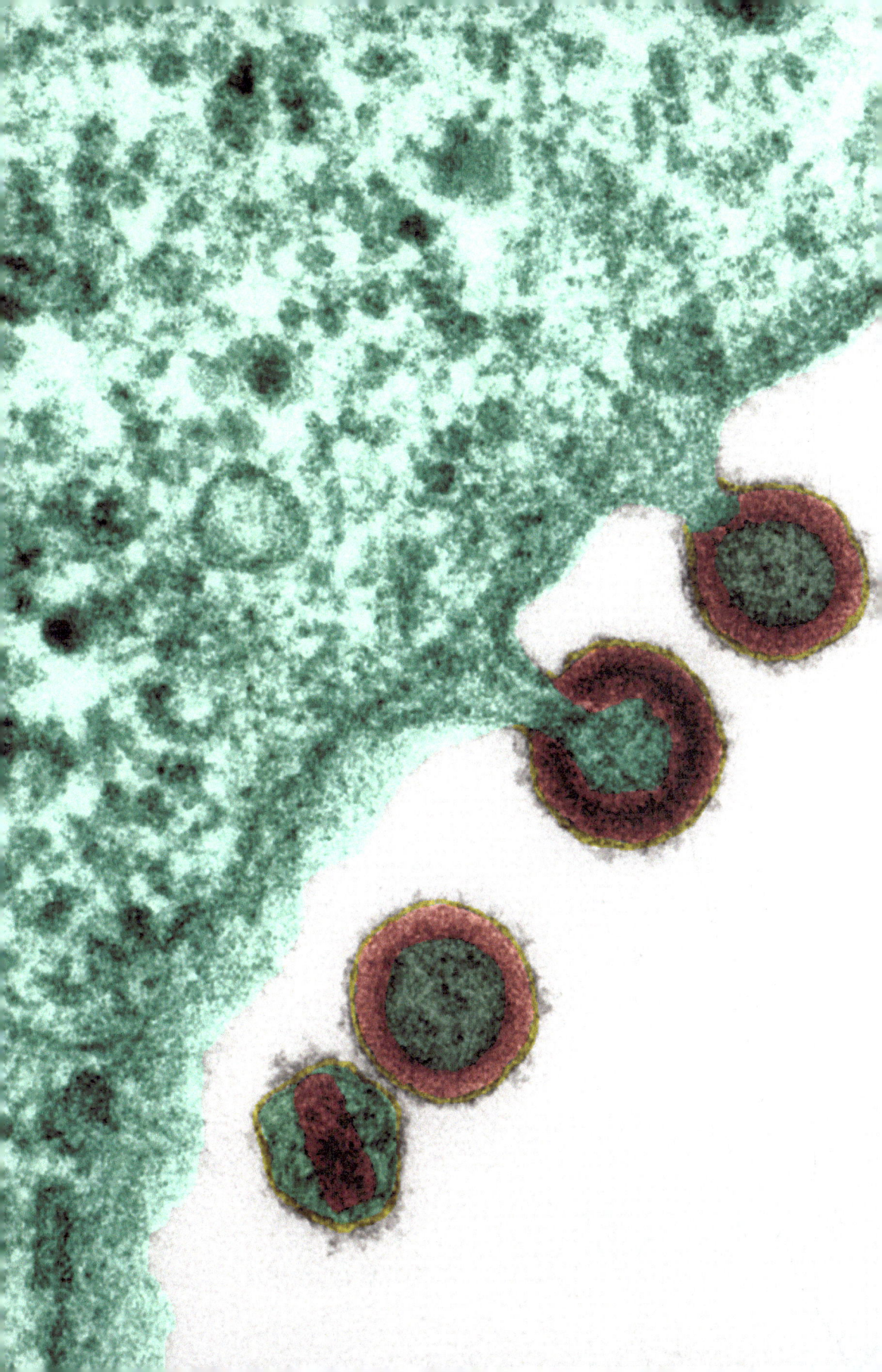

Panel 30. Scanning electron microscopy (SEM) provides another approach to visualize the the HIV budding. The technique does not allow to observe inside the viral particle, particularly the maturation of the **capsid**, but it allows to visualize the entire surface of the viruses and the infected cell. In the case of this photograph, it is thus possible to observe that almost the entire surface of the cell is involved in the budding of viral particles. This illustrates the extraordinary ability of a virus like HIV to multiply in an infected cell, generating a huge number of new viral particles that will be able to infect other cells.

I gave these types of images to the film director Robin Campillo to be used in his film "BPM (Beats per Minute)", "120 battements par minutes" in French, released in 2017. The film tells the story of the AIDS activism of ACT UP, in Paris in the 90s. Robin Campillo's team animated these images to simulate the movement of the viral particle during viral budding and create a visual effect that is used twice during the film.

To date, it is estimated that the 40 million people living with the virus collectively host about 10 million billions **genomes** of HIV (10^{16}). Among all these **genomes**, it is highly likely that there is at least one that is resistant to  the antivirals currently used in clinics, and even to those that will be developed in the future. This should be a reminder that while enormous progress has been made in developing effective antiviral molecules against this virus, we must certainly not relax our vigilance against this virus and its potential future resistance to these antiviral molecules.

HIV
budding

© The University of Tours 2025
P. Roingeard, *Journey to the Viral World: Electron Micrographs of Viruses*,
https://doi.org/10.1007/978-3-031-77995-4_28

actin filaments beneath the plasma membrane. Actin and myosin are the main constituents of muscle cells, responsible for muscle contraction. However, actin and myosin are present in all types of animal cells, but in smaller quantities, to perform important functions of cellular dynamics. Thanks to these proteins, the cell can indeed produce extensions and perform movements.

HIV budding at filopodia

This photograph shows the surface of a cell (in green) where it can be visualized a filamentous surface extension called filopodium. Filopodia are dynamic structures that allow the cell to establish contacts with its environment, i.e. another cell or extracellular matrix on which cells can move. Filopodia are often the site of HIV budding (in blue), as the virus can indeed exploit this cellular extension to its advantage. Its budding at this level is certainly advantageous because filopodia are often in contact with neighboring cells. It allows the virus easier access to new cells to infect, minimizing its exposure in the extracellular environment and to the host's immune system. This process once again illustrates the extraordinary ability of viruses to exploit cellular mechanisms for their benefit.

© The University of Tours 2025
P. Roingeard, *Journey to the Viral World: Electron Micrographs of Viruses*,
https://doi.org/10.1007/978-3-031-77995-4_29

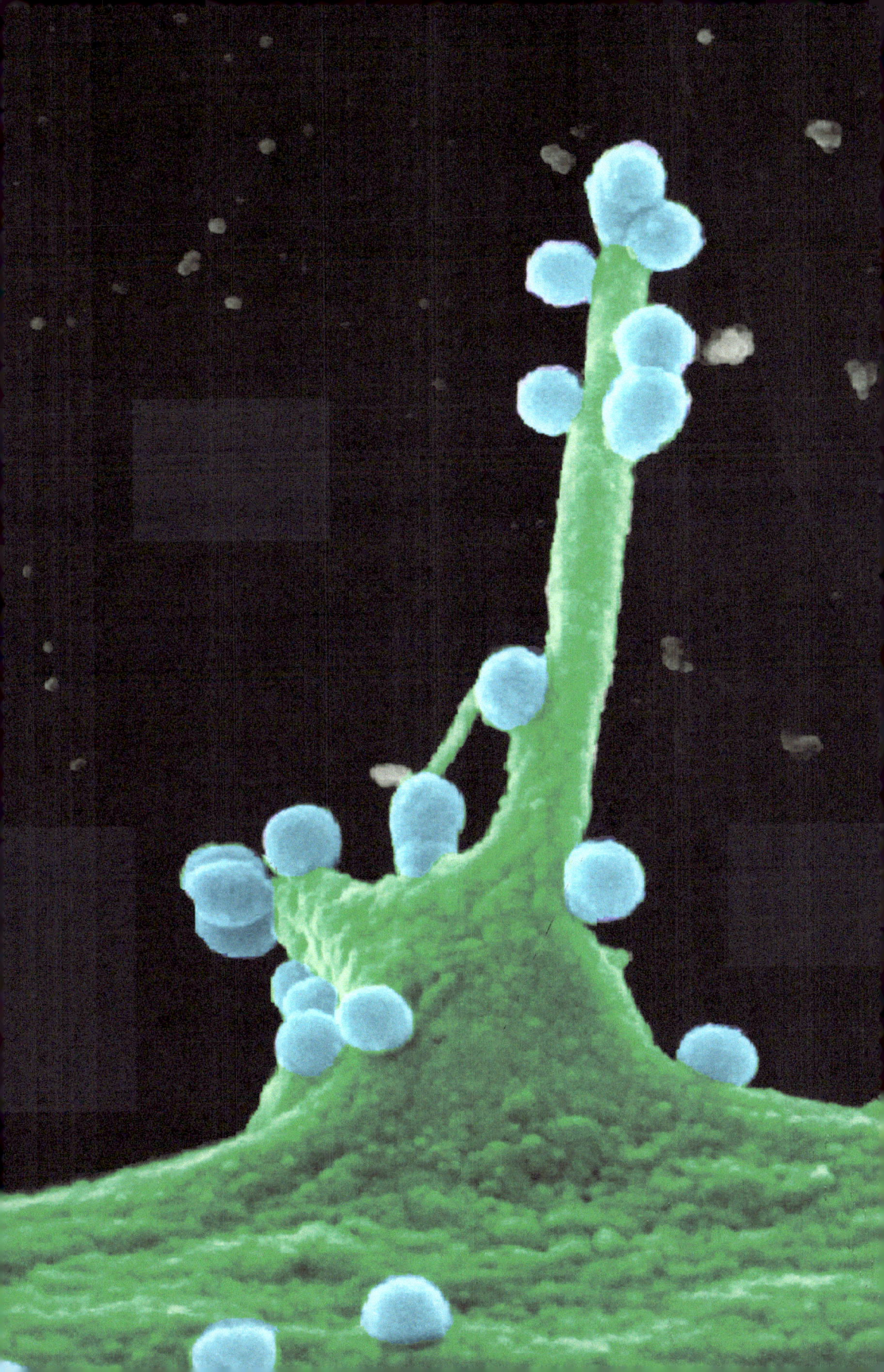

Panel 32. In some cells, HIV does not bud on the surface of the cell, but instead in the cell's internal compartments, as is the case with macrophages. The image at the top shows numerous viral particles present in large internal vacuoles of a macrophage.

Macrophages are the body's "cleaner" cells. They are responsible for phagocytosis mechanisms, meaning they are capable of ingesting cellular debris, even entire cells, bacteria, or other pathogens to digest them. Macrophages can also be infected by HIV. However, unlike T lymphocytes, HIV does not kill them. On the contrary, macrophages can constitute a kind of reservoir, in which HIV can remain for months. Within the intracellular compartments of these macrophages, the viruses are protected from any reaction of the immune system. During this episode, the functions of

the infected macrophages are altered by the presence of the virus, which also contributes to the development of opportunistic infections characteristic of AIDS.

The viral particles present within these intracellular compartments of the macrophages can be released through exocytosis mechanisms, when the cell releases internal constituents to the extracellular space. The image at the bottom shows this mechanism characterized by an opening of the compartment containing the viral particles towards the extracellular environment, following a fusion of the compartment's membrane with the cell's plasma membrane. In consequence, a high number of viral particles are released into the extracellular environment and can then infect many other cells in the body.

© The University of Tours 2025
P. Roingeard, *Journey to the Viral World: Electron Micrographs of Viruses*,
https://doi.org/10.1007/978-3-031-77995-4_30

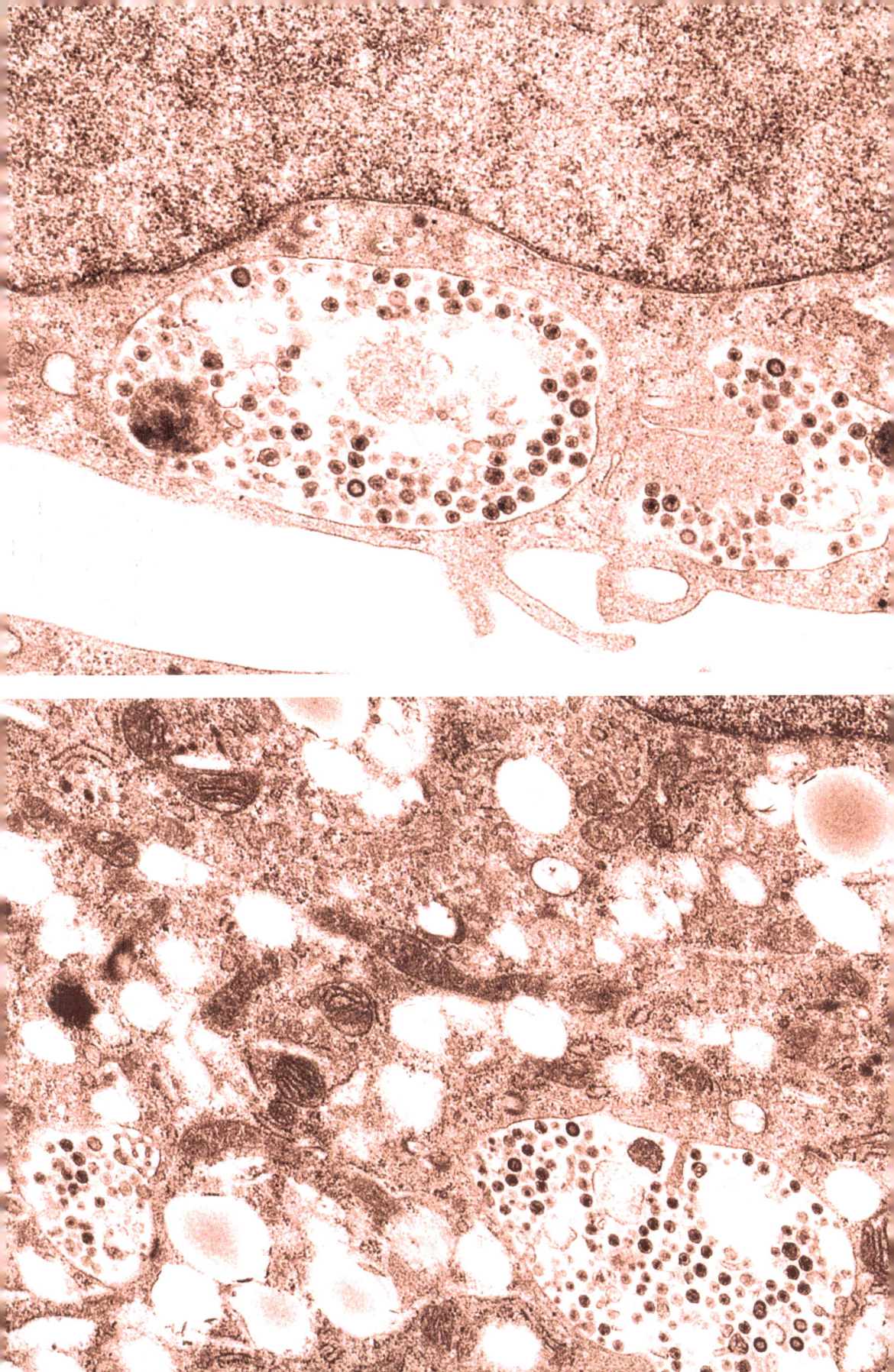

Panel 33. A **syncytium** is a cell formed by the fusion of several cells that share their cytoplasm, but retain their individualized nuclei. The result is a giant cell with multiple nuclei, i.e. a "multi-nucleated" cell. This mechanism exists physiologically to generate multi-nucleated cells of the placenta, bone marrow, or muscle fiber. In the case of HIV, this phenomenon is pathological.

HIV envelope proteins can fuse to the membrane of cells that contain the virus entry receptor, mainly T lymphocytes. A T lymphocyte infected with HIV exhibits the viral envelope proteins on its surface, a preliminary step to the budding of the viral particle at the plasma membrane of the cell. When they reach a certain level of presence on the surface of infected cells, these viral envelope proteins are then capable of recognizing the virus receptors on the surface of other non-infected

HIV induced
syncytium

cells and inducing the fusion of these two cells. The cell issued of this fusion becomes a large infected cell. These mechanisms can then be reproduced with other non-infected cells to lead to the formation of these giant multi-nucleated cells.

The **syncytium** opposite is a cell that has formed by this phenomenon, showing numerous nuclei (in blue). Gigantic cells, carrying several hundred nuclei, can be observed in some infected subjects, particularly in their lymph nodes. It seems that the quantity and size of these giant cells influence the progression of the disease, suggesting that they play a particular role in the pathology and progression towards the AIDS stage. This role, however, is not well known.

Some researchers believe that the **syncytial** cells found in the placenta are formed thanks to genes originating from **retroviruses**. Thus, during evolution, ancestral **retroviruses** could have integrated their **DNA** into cellular chromosomes, then gradually disappear leaving a trace of their passage in the form of certain genes still present in our chromosomes. This would have been beneficial to establish physiological mechanisms related to the presence of these genes. Alternatively, the presence of such genes in cell chromosomes may suggest that viruses appeared during evolution as pieces of nucleic acids that "escaped" from a **genome** cell to become independent.

© The University of Tours 2025
P. Roingeard, *Journey to the Viral World: Electron Micrographs of Viruses*,
https://doi.org/10.1007/978-3-031-77995-4_31

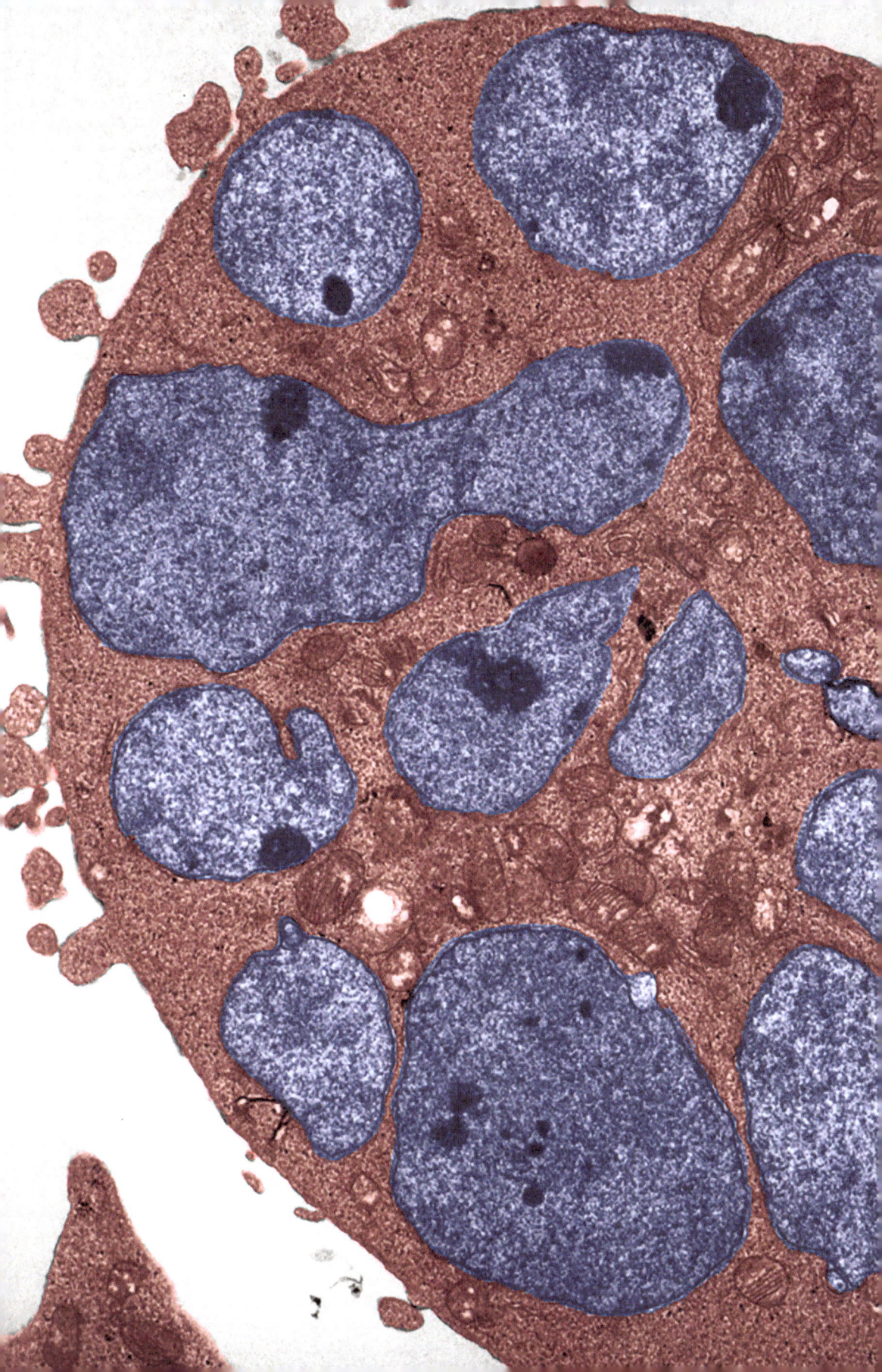

Panel 34. The origin of the name "influenza" is very old and still debated, but it seems that influenza comes from the Italian expression *influenza di freddo* (under the influence of the cold), illustrating the fact that this disease occurs in winter. However, while winter is often the season for people suffering from flu-like syndromes, these syndromes are most of the time benign infections caused by other viruses, such as for example adenoviruses. A real infection by the influenza virus is on the contrary a serious pathology that induces a great fatigue, high fevers coupled with headaches and an acute respiratory syndrome that can be complicated by bacterial superinfections, leading sometimes to pneumonia. It is estimated that this virus causes between 250,000 and 500,000 deaths every year worldwide, especially among the elderly or young children. Human-to-human transmission, during these winter periods when the virus circulates, is done by respiratory route, by drop-

The influenza virus

lets from coughs or sneezes, and probably also by hands. This black and white image obtained by negative staining shows that the virus is quite heterogeneous in shape, ranging from spherical particles of $1/10,000^{th}$ of a millimeter in diameter to more filamentous shapes of $1/10,000^{th}$ of a millimeter in diameter up to $1/1,000^{th}$ of a millimeter in length. It is not visualized in this image, but each viral particle contains 7 or 8 **RNA** segments contained in **capsids** with **helical** symmetry. The crenellated appearance of the surface of the viral particles (especially for the two viral particles colored in the photograph below) is related to the presence of the two envelope proteins of the virus, Hemagglutinin (H) and Neuraminidase (N). There are at least 18 different Hemagglutinins and 11 different Neuraminidases, which can generate 18 x 11 = 198 different combinations, which induces a great variability of the influenza viruses circulating in the nature.

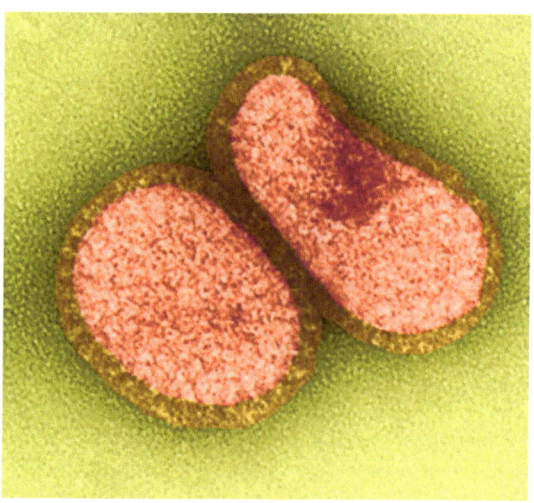

P. Roingeard, *Journey to the Viral World: Electron Micrographs of Viruses*, https://doi.org/10.1007/978-3-031-77995-4_32

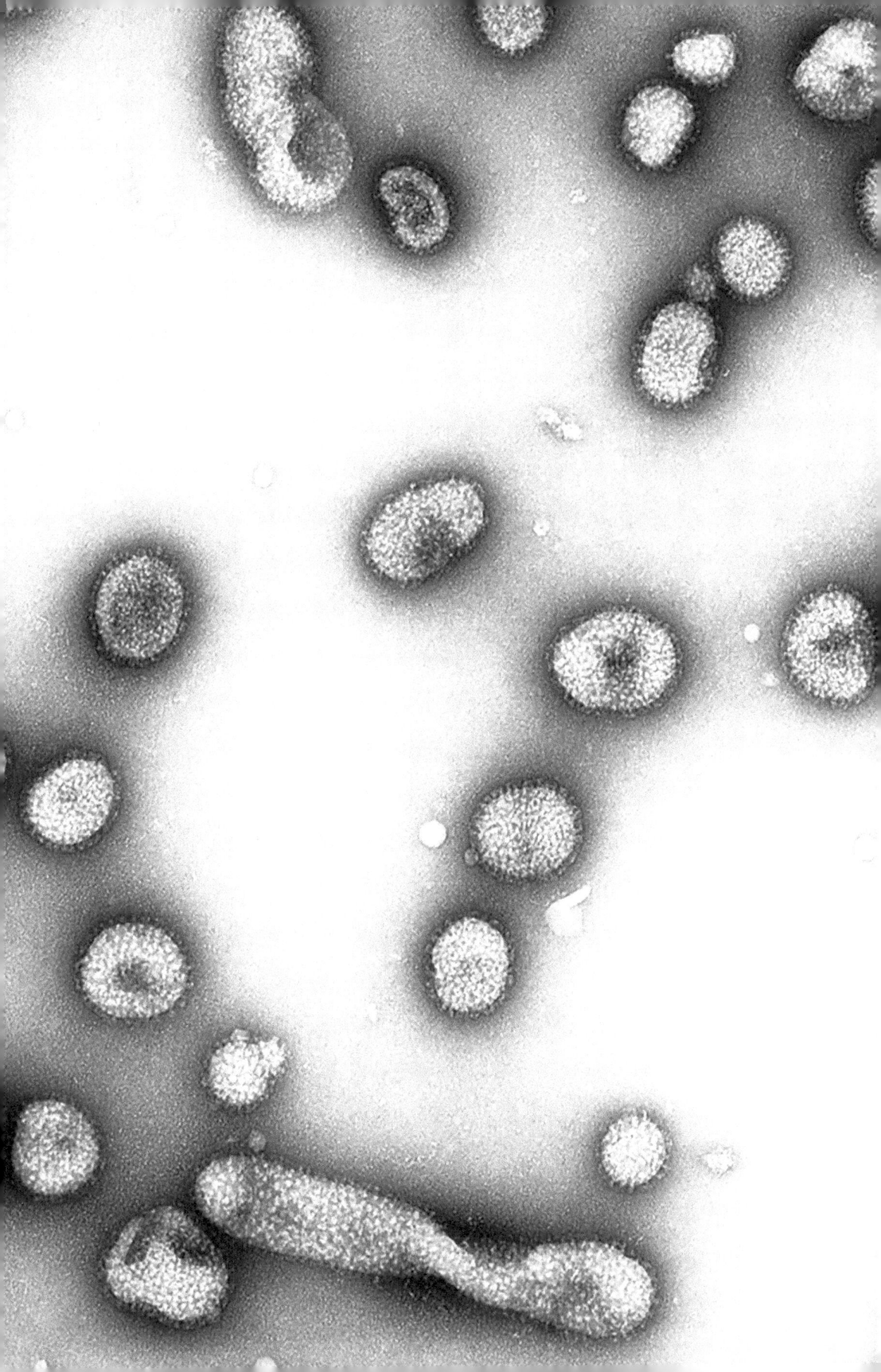

Panel 35. Birds are the reservoirs of influenza viruses, meaning they host these viruses almost permanently, and have done so for a very long time. It is indeed thought that the flu first appeared in birds, 6000 years ago, before infecting humans around 2500 years before Christ, at the time of the domestication of certain bird species by humans. These viruses presumably do not cause pathologies in wild birds, which are therefore "healthy carriers", but they can cause severe symptoms in farmed birds if the viral strain is highly pathogenic. As different viruses can circulate in different bird species and come into contact, recombinations can occur, creating new potentially highly pathogenic viruses responsible for a so-called avian flu. Fortunately, these highly pathogenic avian viruses are generally not transmissible to humans, or, if they are, they lead to a dead end as they cannot spread in the human population via human-to-human transmission. However, pigs are a particularly problematic species from this point of view as they can host both avian and human viruses, which can lead to opportunities for recombinations of highly pathogenic viruses that can easily spread within the human population. This was probably the case during the episode of the so-called "Spanish" flu between 1918 and 1920. Due to a highly virulent strain of an H1N1 virus, this epidemic had resulted in the death of at least 20 million people worldwide. In 2009, a new H1N1 flu epidemic raised significant concern among health authorities. This image shows these H1N1 viruses identified during this winter 2009 epidemic. Fortunately, while this new strain proved to be highly contagious in humans, its **pathogenicity** was not very significant and the mortality associated with the infection was relatively low. As seen in this image of an infected cell section, the viruses have a "hairy" appearance, due to their surface hemagglutinins (H1) and neuraminidases (N1). They appear in the form of rods or spheres, depending on whether the section cut them lengthwise or crosswise.

The H1N1 influenza virus

© The University of Tours 2025
P. Roingeard, *Journey to the Viral World: Electron Micrographs of Viruses*,
https://doi.org/10.1007/978-3-031-77995-4_33

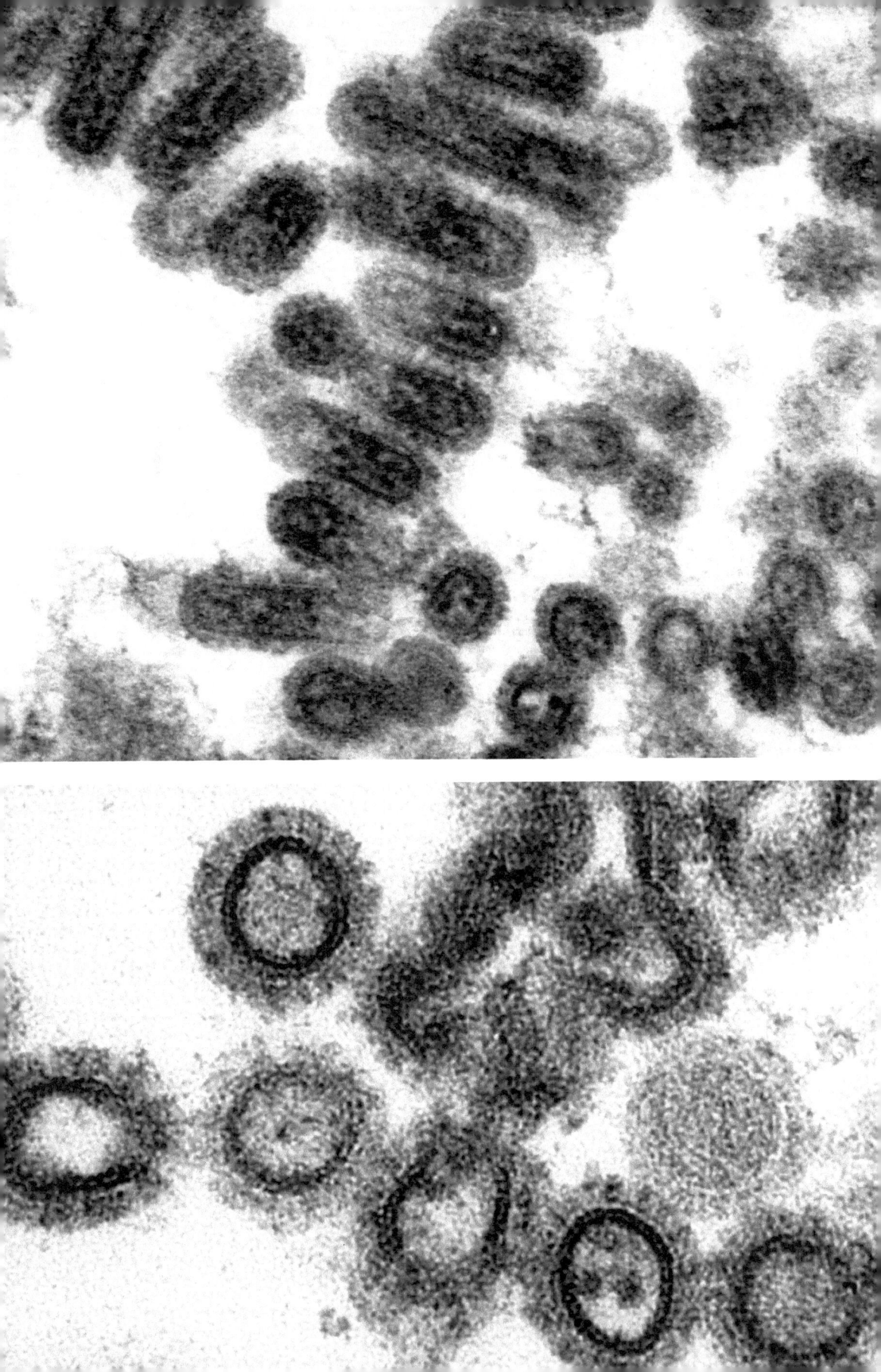

Panel 36. These images obtained by scanning electron microscopy show the influenza virus budding at the surface of an infected cell. The image of the entire cell shows that a large part of its surface is subject to budding of spherical or filamentous viral particles. Even though it is easy to propagate the virus in this way on cell cultures, this method would not be sufficient to generate the amount of virus needed for the production of vaccine batches used every year worldwide. Therefore, a completely different process is used to produce the virus used in the composition of flu vaccines. The virus is inoculated into embryonated chicken eggs, which serve as a real "factory" for virus production; the virus is then purified after extraction from these eggs and chemically inactivated.

Influenza virus budding

The production of a vaccine that must change every year due to the variability of the virus (by combination of different H and N proteins) is also a significant challenge. The World Health Organization (WHO) publishes its recommendations for the composition of vaccines to be used in the northern and southern hemispheres each year. Each of the two hemispheres plays the role of an infectious reservoir for the other hemisphere. Thus, the viruses that circulate in the southern hemisphere during winter are generally those that will infect populations during the northern hemisphere's winter. The risks of poor vaccine efficacy occur in cases of recombination between viral strains, especially if these are strains that have not yet spread widely among populations and for which immunity is weak within populations.

P. Roingeard, *Journey to the Viral World: Electron Micrographs of Viruses*, https://doi.org/10.1007/978-3-031-77995-4_34

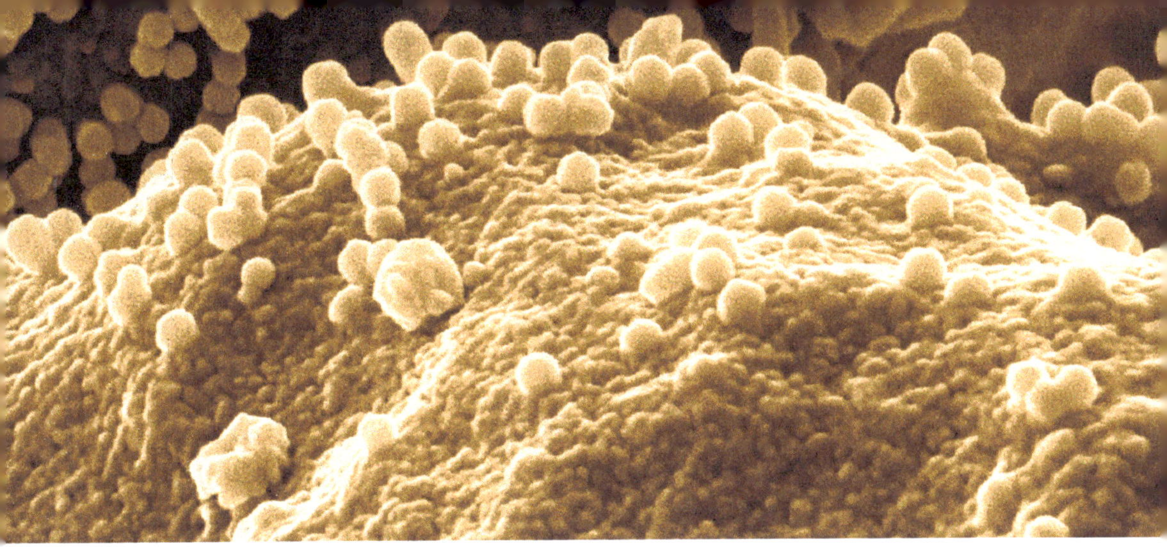

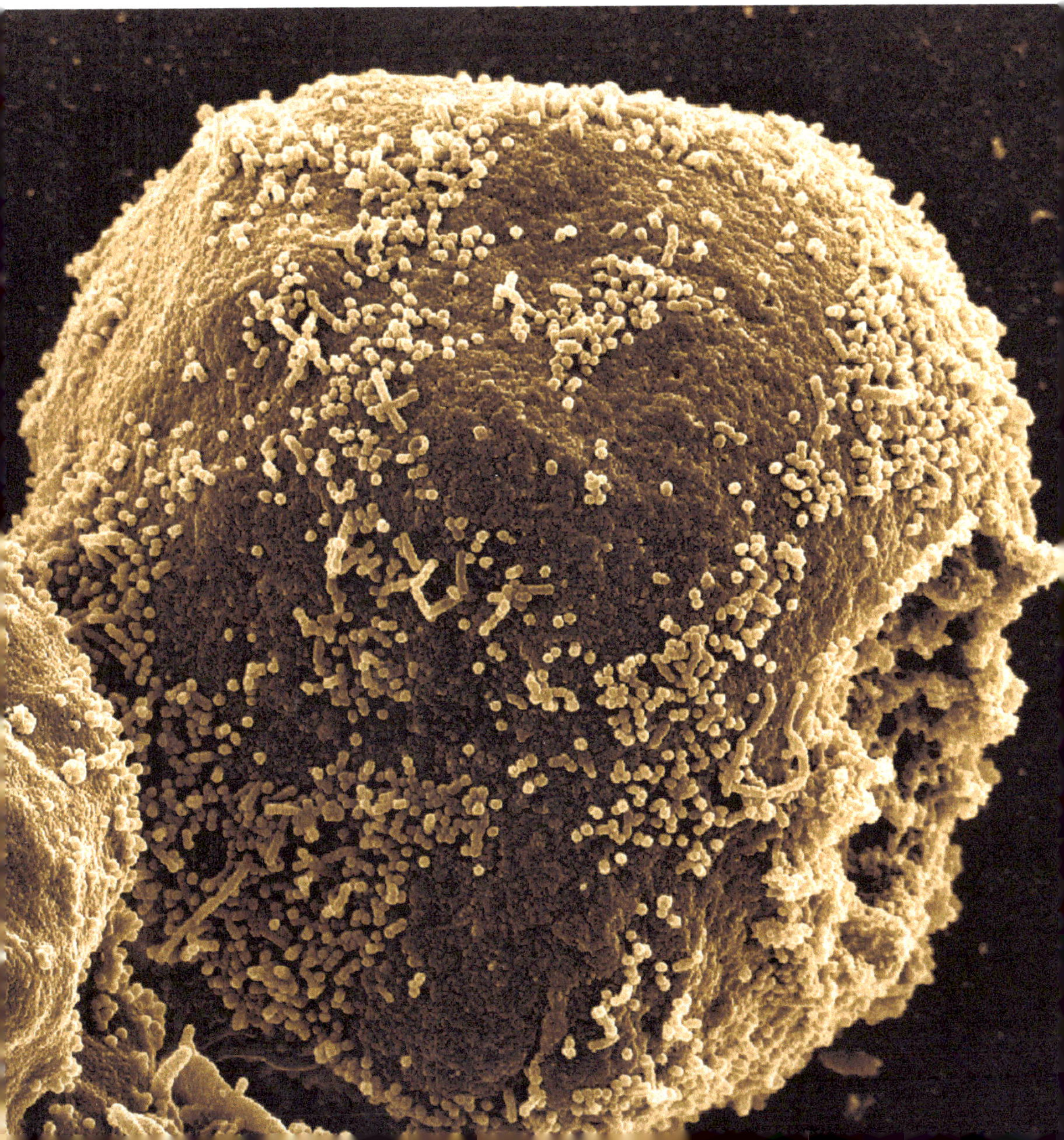

Panel 37. The rabies virus has a very atypical form. It is an enveloped virus that contains a **RNA genome** in a **capsid** with helical symmetry. But its peculiarity is to have a bullet-shaped matrix between its **capsid** and its envelope, as seen in these images obtained by negative staining. The virus is about 1/10000 th in diameter and 1/6000 th in length. The small appendage at the flat base of the bullet is due to the presence of an excess envelope sheath that forms during viral budding on the cell surface.

The rabies virus

This virus infects different species of wild mammals (foxes, wolves, jackals, bats) and transmission to humans is a dead end for the virus, which does not spread through human-to-human transmission. Thus, humans are accidental hosts, usually infected by a bite, the virus being present in the saliva of the infected animal. This bite can occur directly from an infected wild mammal, but more commonly through a domestic animal like a dog, itself bitten by a wild animal or another infected dog. There are cases of transmission by certain species of bats that can be healthy carriers, and therefore reservoirs of this virus. However, these cases are very rare. The rabies virus infects nerve cells and induces encephalitis that leads to death in the absence of treatment.

© The University of Tours 2025
P. Roingeard, *Journey to the Viral World: Electron Micrographs of Viruses*,
https://doi.org/10.1007/978-3-031-77995-4_35

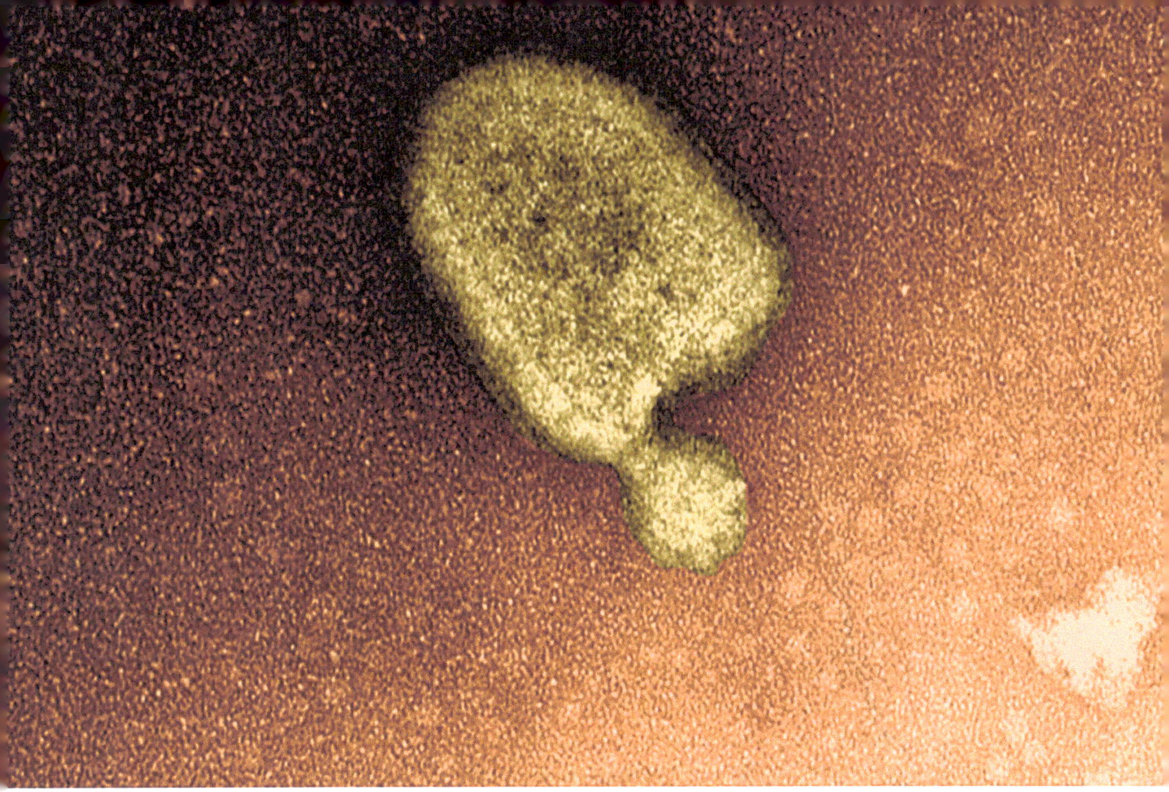

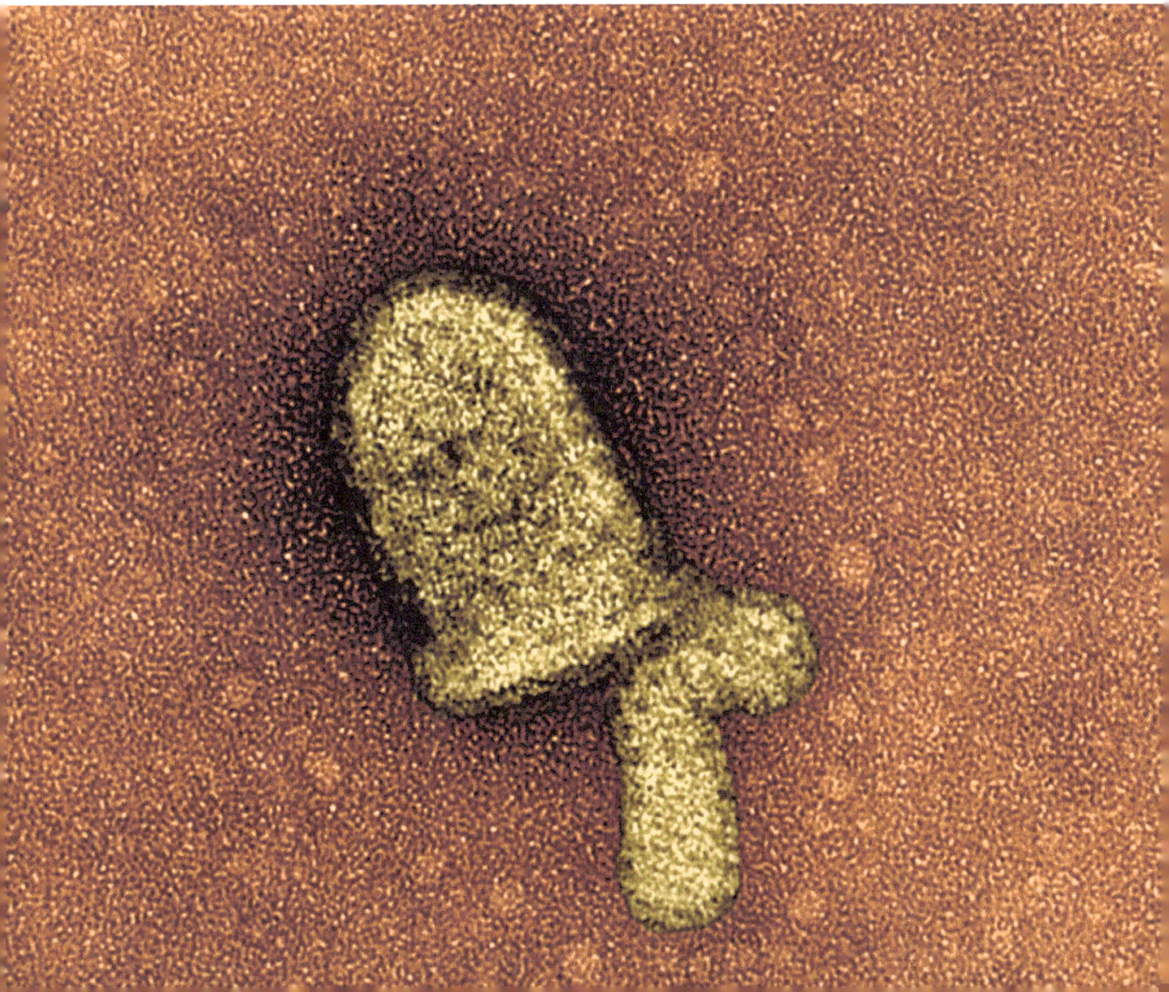

Panel 38. These photographs of cells infected by the rabies virus show the virus budding at the plasma membrane of these cells. The bullet-shaped matrix is surrounded by a viral envelope. This viral envelope is made up of a part of the plasma membrane borrowed from the host cell, in which the viral envelope proteins are embedded. These envelope proteins give, here also, a crenellated appearance of the viral surface. As these are cell sections, the bullet-like aspect is well visualized when the virus is cut lengthwise; the virus appears circular when cut transversely.

During an infection, the first cells infected by the virus are the cells present at the bite site, and therefore the cells of the surrounding muscle tissues. The virus then gradually reaches the brain via the nerve endings. This can take a few days or a few weeks, depending on the amount of virus present in the saliva of the infected animal and the location of the bite (the closer it is to the brain, the faster the virus will reach it). These considerations are very important as they determine the chances of success of a vaccination. The vaccine must be administered as soon as possible so that the body can

Rabies
virus budding

defend itself against the virus before it reaches the brain. Without this intervention, rabies is always a fatal disease. The virus replicates in the brain, then reaches the salivary glands, the eyes. It induces behavioral disorders, aggression, and the infected subject eventually dies from loss of vital functions. The first successful rabies vaccination was conducted in France by Louis Pasteur, in 1885, on a 9-year-old child who had been bitten by a rabid dog. The rabies vaccination at the time involved administering an attenuated virus, i.e. a virus that has all the characteristics of the normal virus except its **pathogenicity**. This attenuated virus therefore stimulated the immune system, but without inducing pathology. Modern vaccines are made up of inactivated viruses, i.e. pathogenic viruses, but killed by heat.

© The University of Tours 2025
P. Roingeard, *Journey to the Viral World: Electron Micrographs of Viruses*,
https://doi.org/10.1007/978-3-031-77995-4_36

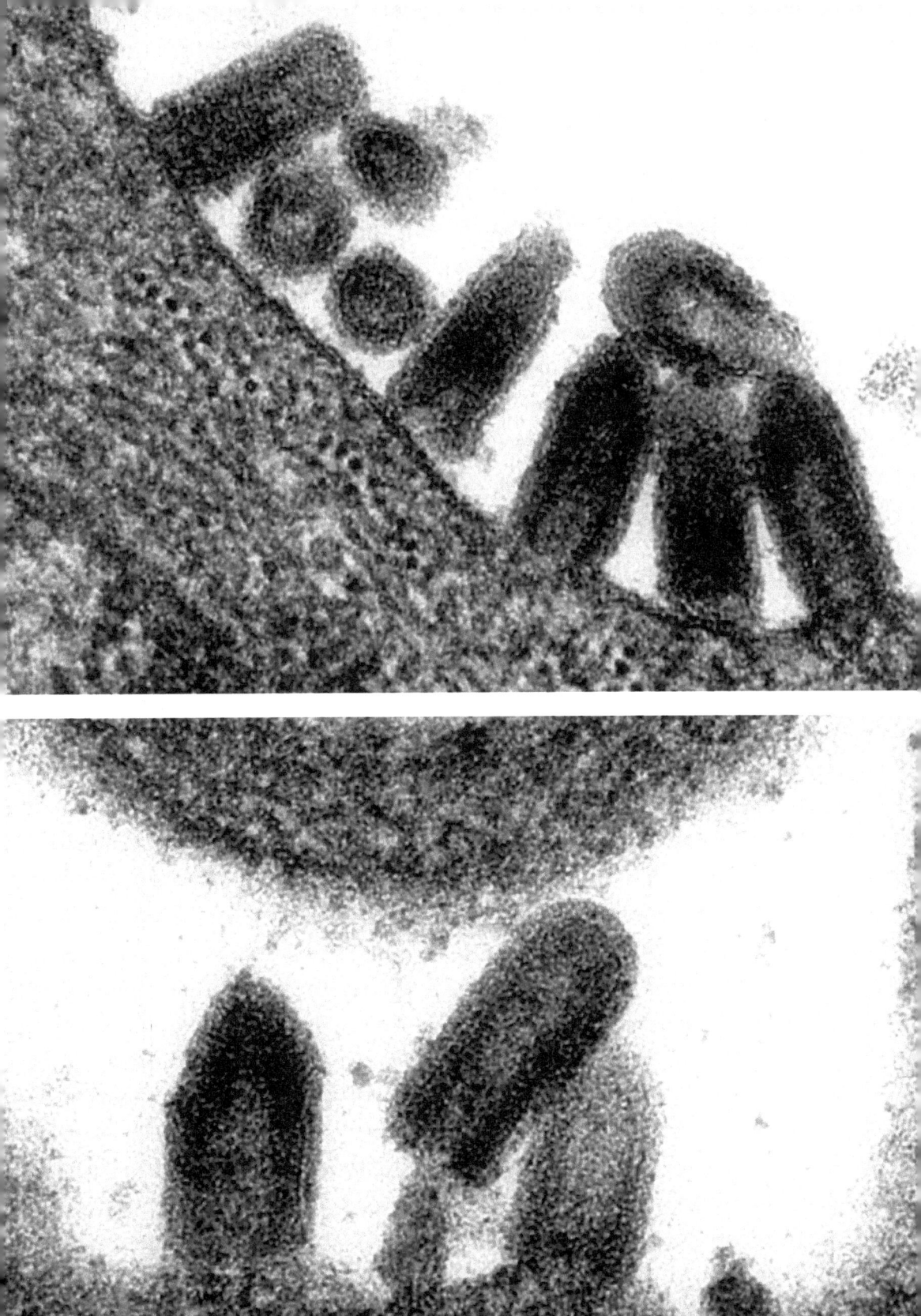

The rabies virus

on a cell surface

Panel 39. This photograph of a cell infected by the rabies virus, obtained by scanning electron microscopy, shows the virus budding at the plasma membrane. This visualization clearly shows this particular shape of viruses that point at the surface of cells, which here evoke images of war missiles.

It is estimated that this virus causes about 50,000 deaths per year worldwide, mainly in rural areas of Africa and Asia, where dogs are the main reservoir of the virus and therefore the main source of human contamination. In industrialized countries, the rabies virus only persists in wildlife. In Europe, dog rabies has been eradicated for several decades, and the virus adapted to wildlife at the end of the Second World War, mainly in foxes and raccoon dogs.

The virus is no longer present at all in certain territories, either due to a particularly favorable geographical situation, such as Australia or UK, or by its elimination through oral vaccination programs, as in the countries of Central and Western Europe. However, there is still a residual risk of the occurrence of rabies cases in these countries, coming from illegally imported animals.

© The University of Tours 2025
P. Roingeard, *Journey to the Viral World: Electron Micrographs of Viruses*,
https://doi.org/10.1007/978-3-031-77995-4_37

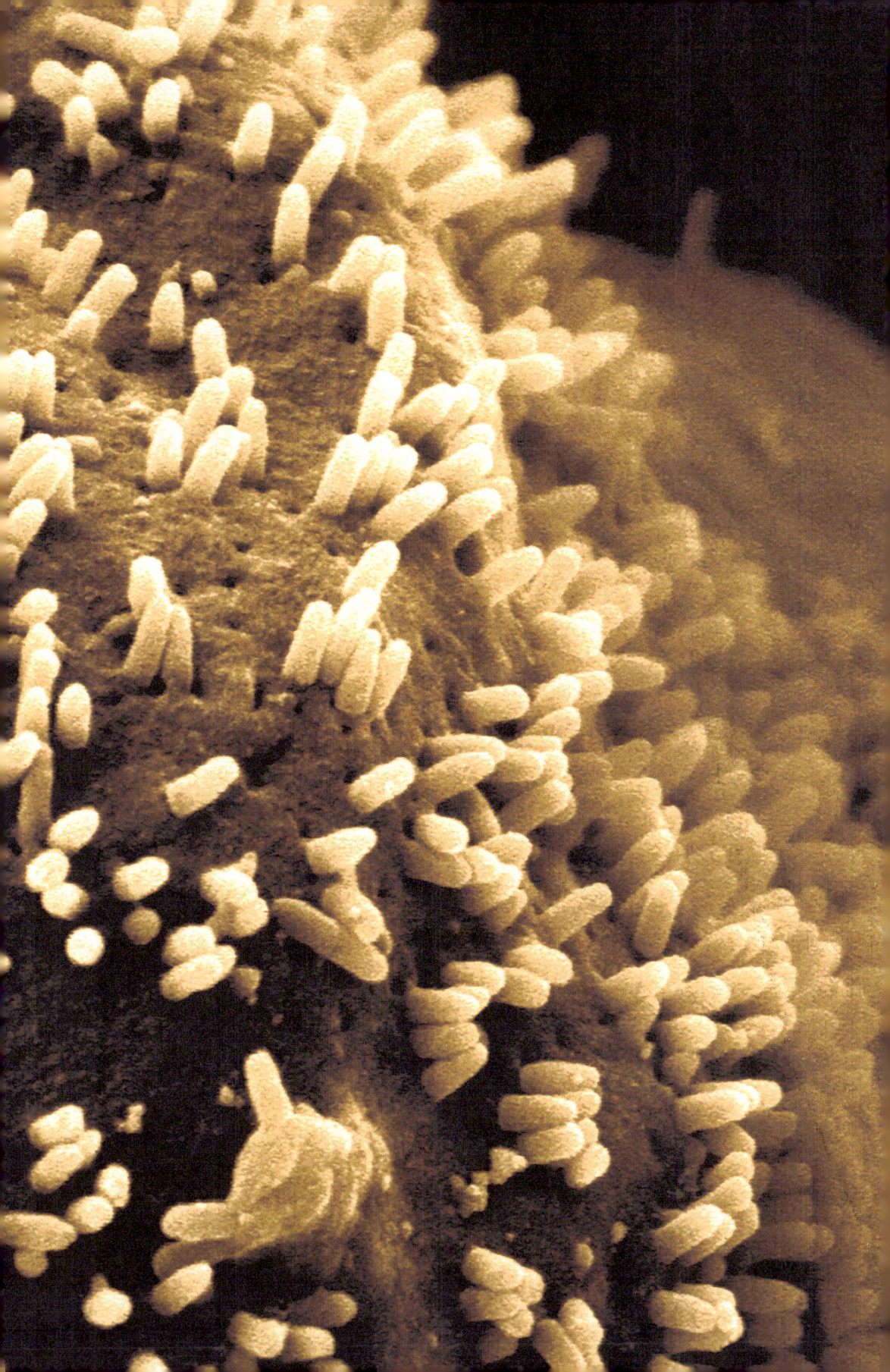

Panel 40. The measles virus is a large enveloped virus with a **capsid** of **helical** symmetry. The virus is more or less spherical, with a heterogeneous size (from $1/10000^{th}$ to $1/8000^{th}$ of a millimeter). This photograh shows measles virus produced by an infected cell, with their envelope (in yellow), surrounding a matrix (in dark green). The **capsid**, inside this matrix, is not distinguishable in this image. Some areas of the cell membrane show an accumulation of matrix and envelope proteins, representing the events that precede viral budding, just before the formation of the viral particle. The measles virus is a strictly human virus that is extremely contagious. It causes one of the eruptive diseases of early childhood. The virus always enters through the respiratory route. The child or young adult comes into contact with saliva droplets (coughs, sneezes). The virus then circulates in the body and causes typical skin eruptions and multiple other clinical signs (fever, headaches, cough…). In the vast majority of cases, the immune system eliminates the viruses, but there are nevertheless very serious complications of this infection, when the virus infects the lungs or the brain. Some can be fatal.

There is no treatment against this virus, only vaccination can prevent infection. Fortunately, an effective vaccine is available, which is estimated to have prevented more than 60 million deaths worldwide between 2000 and 2023.

The measles virus

© The University of Tours 2025
P. Roingeard, *Journey to the Viral World: Electron Micrographs of Viruses*,
https://doi.org/10.1007/978-3-031-77995-4_38

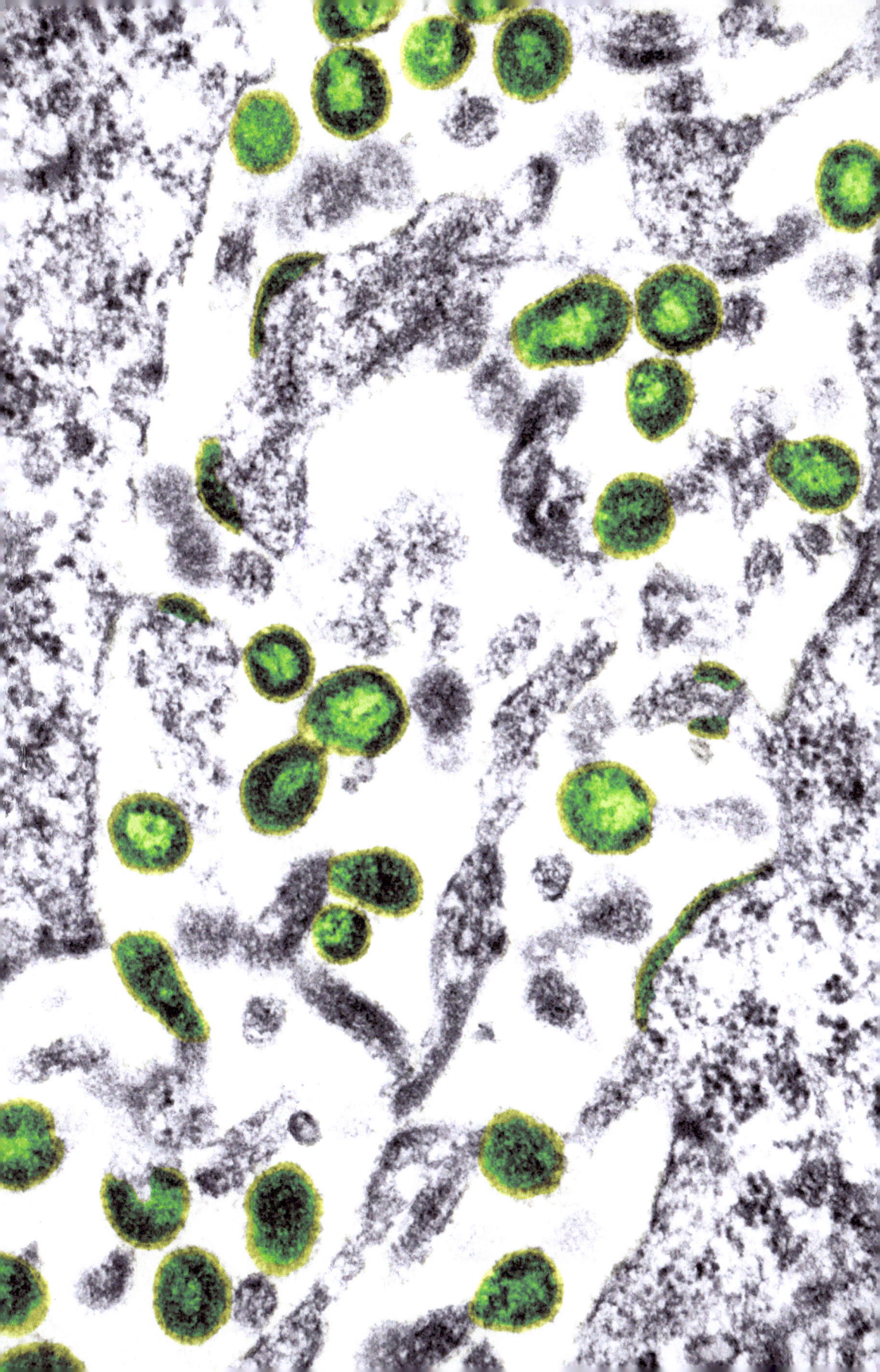

Panel 41. This photograph obtained by scanning electron microscopy shows measles viruses budding at the surface of an infected cell. The virus is formed by borrowing a portion of membrane from the cell it infects before individualizing and being released into the extracellular environment.

The measles vaccine is made using cells that produce such viruses, which is then attenuated. The vaccine is therefore able to stimulate the immune system to induce protection without causing disease. This type of attenuated virus is obtained by replicating the virus in cells of another species. After many passages, the virus continues to multiply, but gradually loses its virulence.

Measles virus budding

This vaccine has been available since the 1970s. It is now offered in the form of a vaccine combining attenuated viruses against mumps and rubella (MMR vaccine for "Measles-Mumps-Rubella"). It is estimated that the virus causes about 100,000 deaths per year worldwide, mainly in children under 5 years of age, due to insufficient vaccination coverage.

© The University of Tours 2025
P. Roingeard, *Journey to the Viral World: Electron Micrographs of Viruses*,
https://doi.org/10.1007/978-3-031-77995-4_39

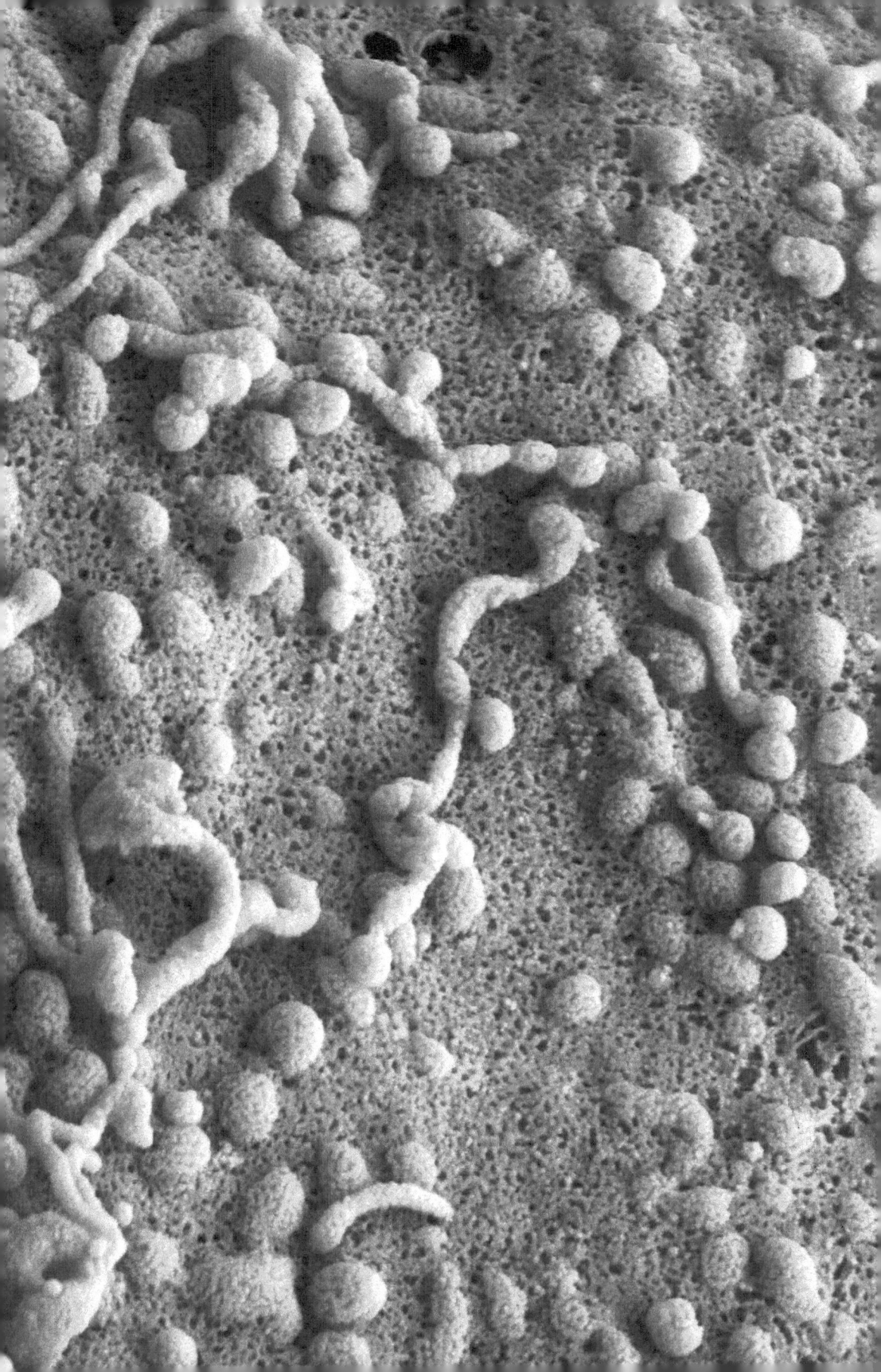

Panel 42. The mumps virus is a large, enveloped virus, more or less spherical, about 1/8000th of a millimeter in diameter. Its envelope is relatively fragile. The photograph at the bottom shows a burst virus, revealing its **capsid** with **helical** symmetry. This virus infects the salivary glands located between the upper jaw and the ears. The inflammation of these glands causes ear pain and difficulty opening the mouth to talk or eat. These symptoms probably gave the illness its name, as it could come from "mump", an old English word for grimace.

The virus is found in large quantities in saliva of infected people and is thus very contagious. It is transmitted through the air, by droplets of saliva (coughing, sneezing, spitting…). It causes headaches and fever, but the infection is generally benign, well controlled by the immune system, and therefore without consequence in children. However, it can induce serious complications in young adults, including testicular damage that can lead to sterility. This infection, which was described as early as antiquity, can be well controlled thanks to vaccination, particularly through the use of the MMR vaccine (see panel 41), which contains an attenuated mumps virus.

The
Mumps virus

© The University of Tours 2025
P. Roingeard, *Journey to the Viral World: Electron Micrographs of Viruses,*
https://doi.org/10.1007/978-3-031-77995-4_40

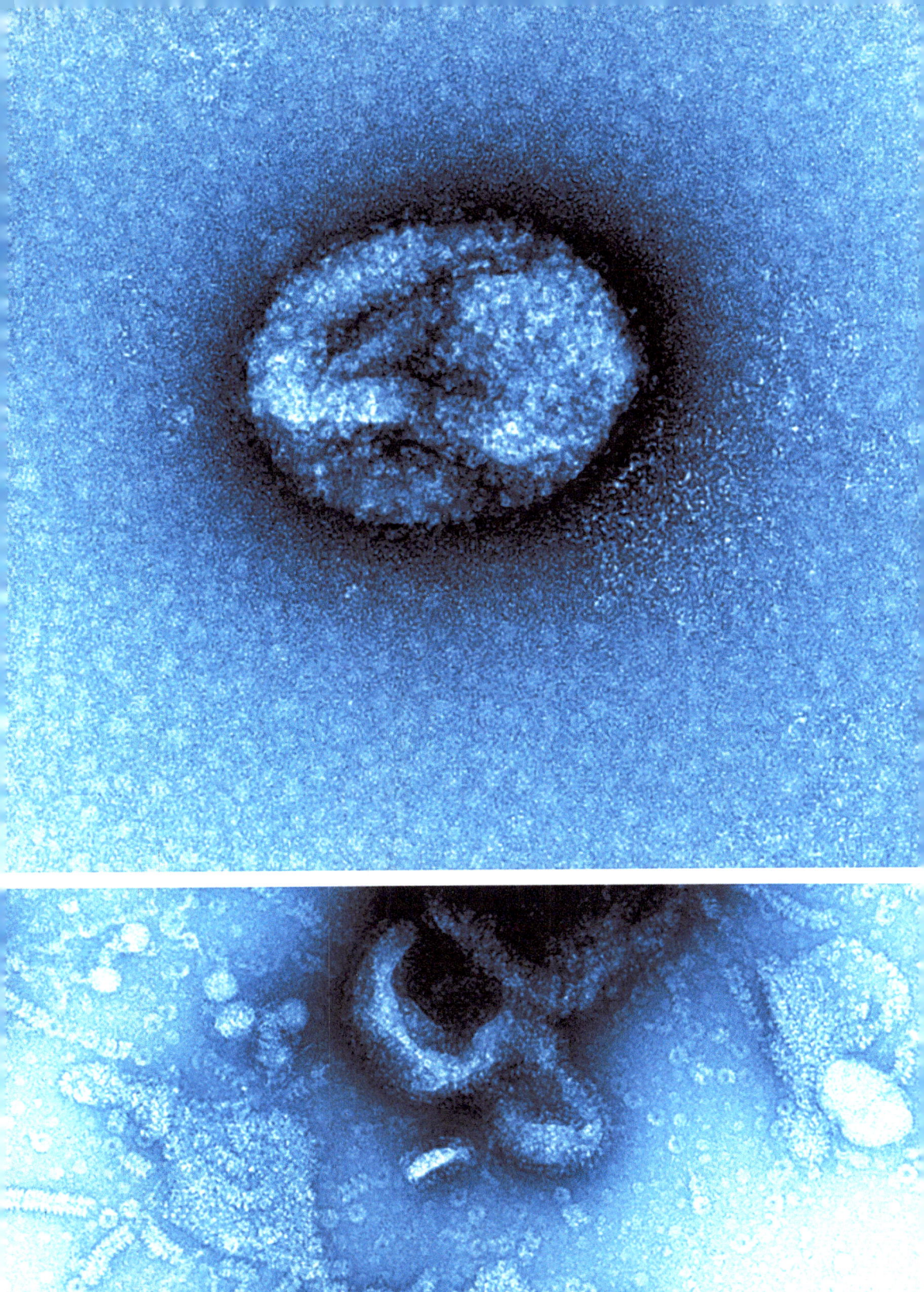

Panel 43. The rubella virus is a small enveloped virus (about 1/50000 th of a millimeter in diameter), containing an **icosahedral capsid**. In this photograph obtained by negative staining, the envelope has been colored in purple, and the **capsid** colored in green. The **capsid** is not visible in all viruses, only in those that have been slightly permeabilized during deposition on the electron microscopy grid, allowing the metal salt to penetrate the particle. This virus is strictly human. It induces a benign infection, characterized by a skin rash accompanied by a fever and headaches. The name rubella is derived from Latin, meaning "little red." Also transmitted by airborne route and highly contagious, rubella has long been confused with measles, giving the same clinical signs. Rubella was first described as a separate disease in the German medical literature, in 1814, and for this reason is also know as the "German measles." However, it is a completely different virus. If the infection by this virus is benign, it is associated with serious risks of congenital

The rubella virus

diseases in newborns, especially if the infection occurs early in pregnancy. The virus, which initially multiplies in the respiratory mucosa and then in the lymph nodes, can, via the blood, contaminate the fetus and cause serious lesions. It can thus disrupt the fetal development process, inducing different types of malformations.

Due to systematic vaccination of children with an attenuated virus (it is one of the 3 attenuated viruses included in the MMR vaccine composition, see also panel 41), congenital rubella has practically disappeared in many countries. Unfortunately, it still exists in countries where vaccine coverage is not as good.

© The University of Tours 2025
P. Roingeard, *Journey to the Viral World: Electron Micrographs of Viruses,*
https://doi.org/10.1007/978-3-031-77995-4_41

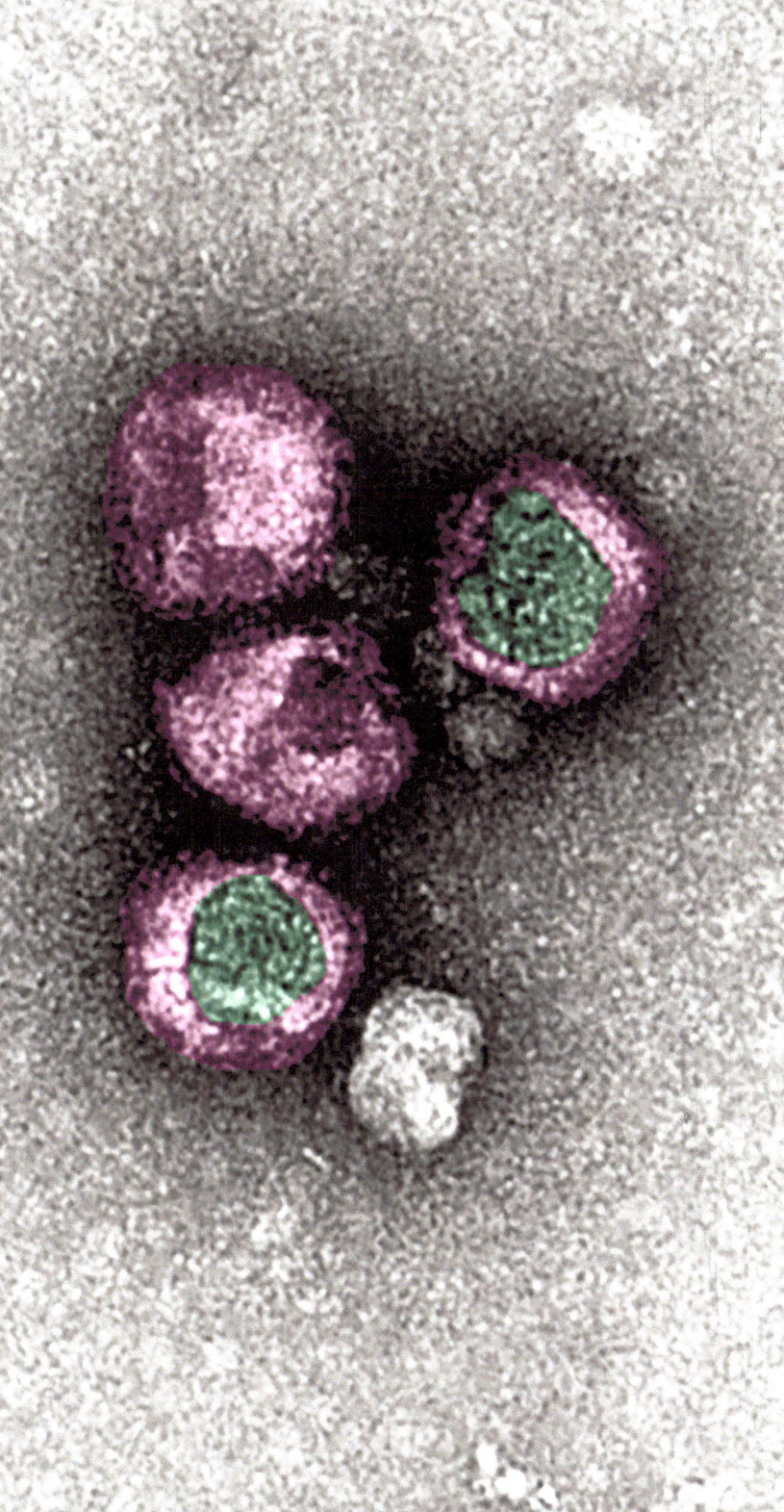

Panel 44. This name of poxviruses comes from the ability of these viruses to induce skin pustules (as pox is the plural of pock, a pustule). This photograph shows the negative staining of poxviruses inducing an infection named *molluscum contagiosum*. In solution, the virus resembles a large brick about 1/4000 th of a millimeter long by 1/6000 th of a millimeter wide. It induces pustules on the skin containing many copies of the virus. Human-to-human transmission occurs through skin-to-skin contact or via towels. These infections are benign as they regress spontaneously, without any particular intervention. It is just recommended to protect the lesions to avoid contagion because, as the name of the infection indicates, this virus is extremely contagious.

Pox-viruses

However, humans have long been the target of an infection by a poxvirus responsible for very serious and highly contagious infections with respect to human-to-human transmission: the smallpox virus. This virus has existed for millennia since traces of infection by the smallpox virus have been detected on Egyptian mummies. The virus caused thousands of deaths until the 18 th century when the first vaccinations began to curb its spread. Mass vaccination campaigns were finally able to lead to a complete eradication of the virus from the surface of the planet, in 1977.

In a very interesting way, it is an animal virus from the same family, present in cows, which allowed the development of the smallpox vaccine by Edward Jenner. This virus, initially called "vaccine" virus, gave its name to the concept of a vaccine. This cowpox virus is not very pathogenic in humans and is able to induce a protective immune response against the smallpox virus.

The English terminology around poxviruses uses the term *smallpox* to designate the smallpox virus, *cowpox* for the cow's poxvirus. But curiously, this same terminology uses the term *chickenpox* to designate the varicella virus (also see panel 22) even though this disease is caused by a strictly human herpes virus that does not infect birds. The origin of the word is strange and not known with certainty; but this name of *chickenpox* could come from the fact that varicella also causes skin eruptions, but with much less serious consequences (when it circulated in populations, the smallpox virus indeed killed about one in five patients). The word *chicken* also means "coward" in popular language, and this term was possibly used to mark the much less dangerous nature of chickenpox compared to smallpox

© The University of Tours 2025
P. Roingeard, *Journey to the Viral World: Electron Micrographs of Viruses*,
https://doi.org/10.1007/978-3-031-77995-4_42

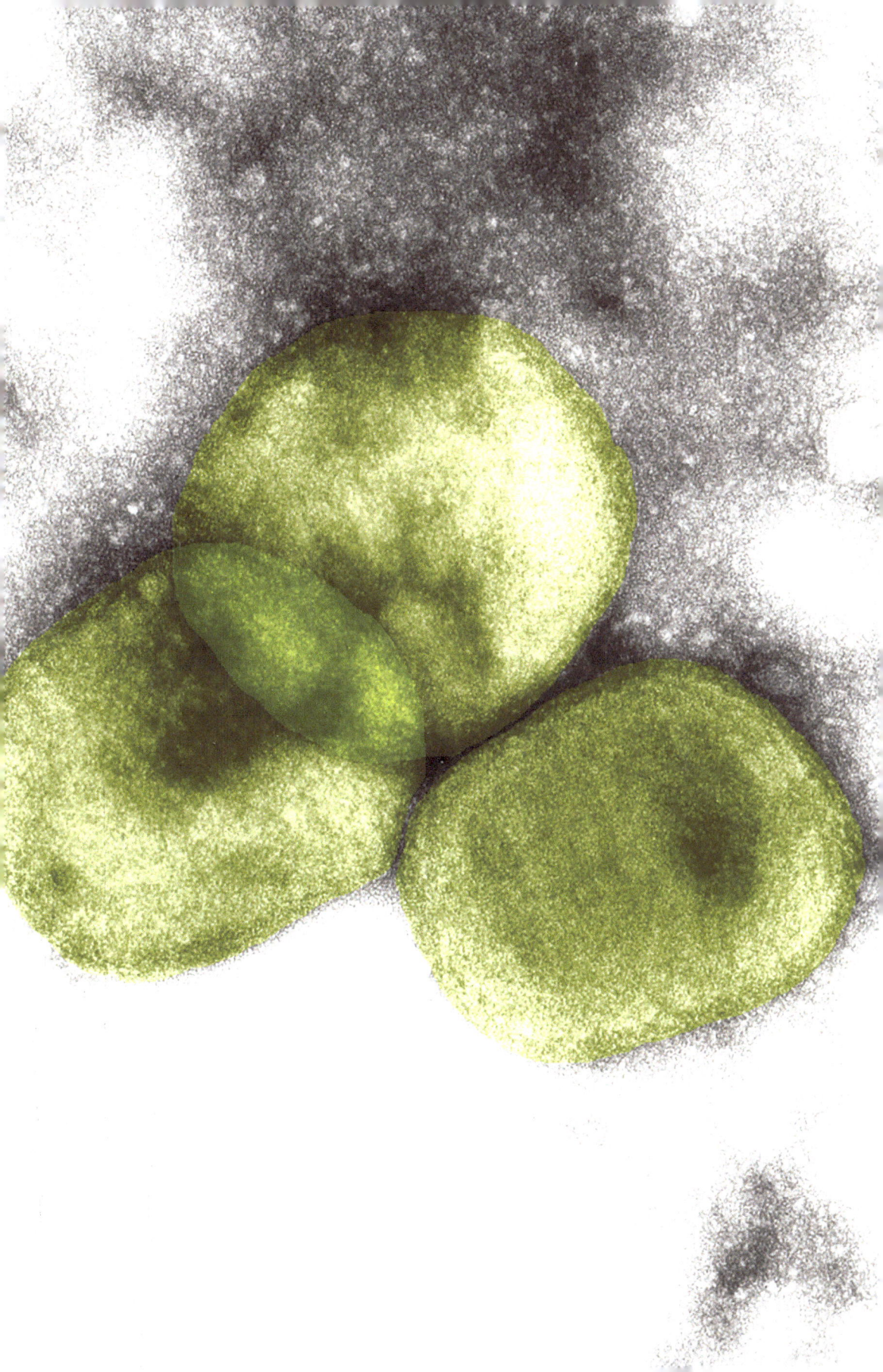

Panel 45. The orf virus, visualized here in a skin section, is part of the large family of poxviruses. Poxviruses are indeed viruses that infect many animal species, far beyond humans and cattle: monkeys, deer, pigs, squirrels, crocodiles, sheep, birds, and even insects. They are

The Orf
Virus

large **DNA** viruses, enveloped, possessing an unusual **capsid** in the world of viruses, as this capsid is oval-shaped. As a result, the virus is also more or less oval-shaped, as shown on this photograph, on which the **capsid** has been colored in red, and the envelope in green. The spaces between the **capsid** and the envelope, colored here in brown, are called lateral bodies.

The orf virus mainly infects sheep and goats. It is very easily transmitted between animals in farms and can contaminate humans through contact with skin lesions of an infected animal. It then induces skin eruptions on the hands or face of the infected person. This strictly animal virus cannot then spread by human-to-human transmission. Its transmission from animal to human is therefore a dead end for the virus.

The orf virus shown in this image was observed in pustular lesions present on the hands of a person who worked in a slaughterhouse and who was in contact with sheep. These lesions are painful, but relatively benign as they spontaneously regress within a few weeks. In the 19 th century, it was common to observe pustules similar to those induced by the orf virus on the hands of dairy farmers who were hand-milking. These pustules were due to the cowpox virus that was present on the cow udders. They were referred to as "milker's nodules" to describe these benign lesions. They have almost disappeared with the introduction of automatic milking machines.

© The University of Tours 2025
P. Roingeard, *Journey to the Viral World: Electron Micrographs of Viruses*,
https://doi.org/10.1007/978-3-031-77995-4_43

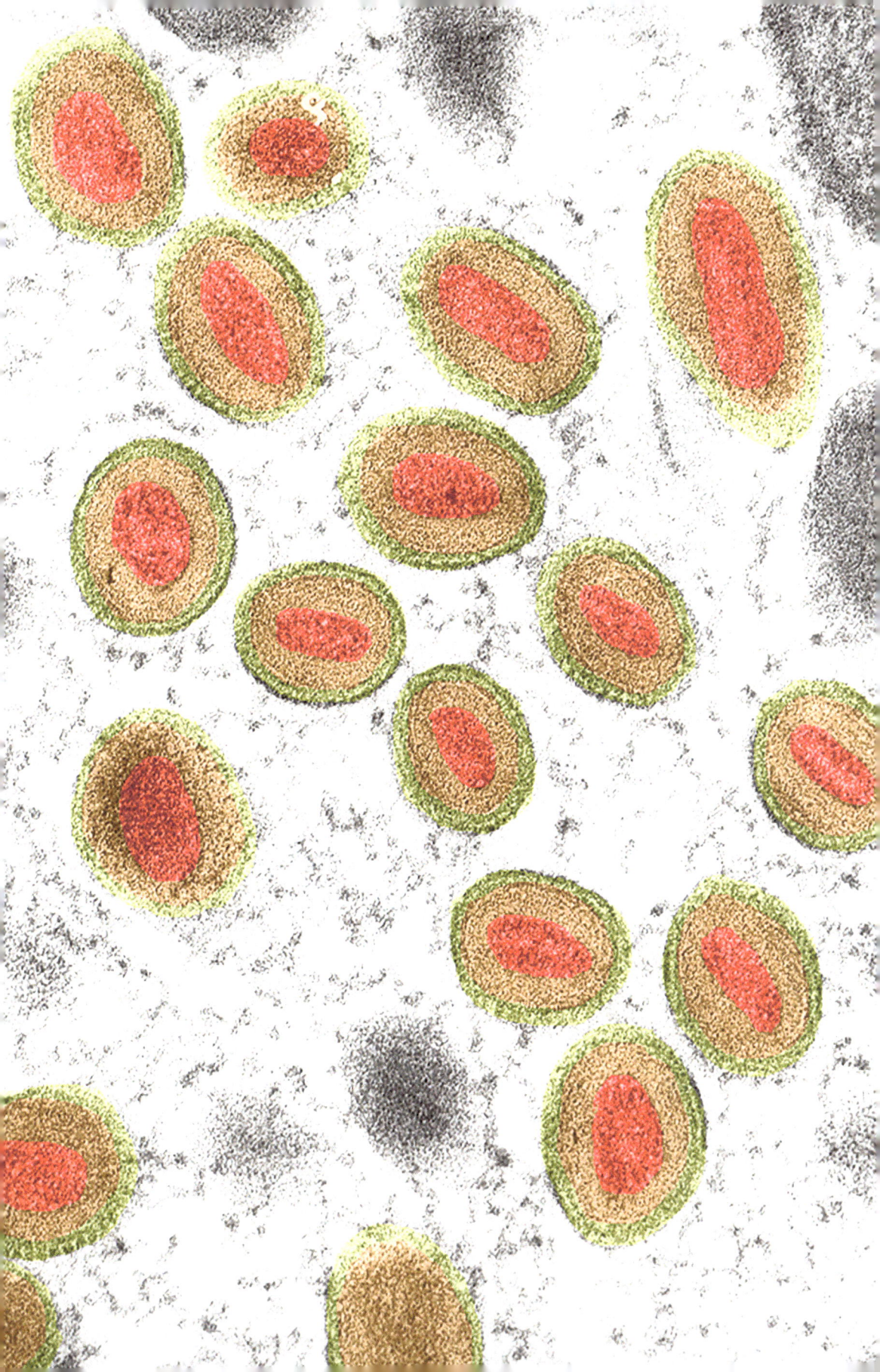

Panel 46. There are 5 hepatitis viruses, simply called A, B, C, D, and E. The two most concerning viruses in terms of public health are the hepatitis B (HBV) and hepatitis C (HCV) as they chronically infect a very large number of people worldwide (respectively 250 and 70 million). These chronic infections can evolve over the long term into serious liver diseases, even leading to the development of liver cancer.

HBV is a very peculiar virus because it circulates in the blood of infected subjects accompanied by excess envelope particles. The small photographs bellow show the 3 types of viral particles that are present in the blood of HBV chronic carriers. The photograph on the left shows the virus itself, about 1/50000 th of a millimeter, consisting of a spherical **capsid** (orange), containing a **DNA genome**, which is surrounded by an envelope (in yellow). The photographs in the middle and on

The Hepatis B virus

the right show particles that only contain the envelope, and therefore lack **capsid** and **DNA**. These excess envelope particles or "subviral" particles form small spheres of 1/80000 th in diameter or filaments of 1/80000 th wide and of variable length, which are therefore non-infectious. As seen in the opposite photograph, these subviral particles are, in the blood of infected subjects, in large excess compared to the virus itself. Their supposed role is to divert the antibodies produced by the immune system to the benefit of the virus. As they are in large excess, these subviral particles would constitute a "decoy" for the antibodies, which would facilitate the virus's escape from the immune system.

This observation gave Professor Philippe Maupas and other scientits the idea to create a vaccine against HBV from these excess envelope particles. Developed in Tours at the end of the 1970s, a first-generation vaccine was based on the purification of these excess envelope particles from the blood of subjects chronically infected with the hepatitis B virus. As these particles lack a capsid, and therefore DNA, they are non-infectious and can thus fulfill this role of stimulating the immune system to induce protection, without being able to induce the disease.

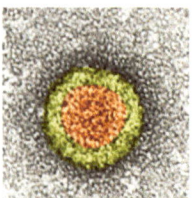 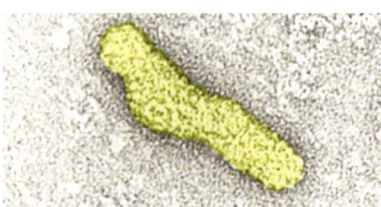 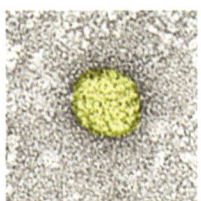

P. Roingeard, *Journey to the Viral World: Electron Micrographs of Viruses*,
https://doi.org/10.1007/978-3-031-77995-4_44

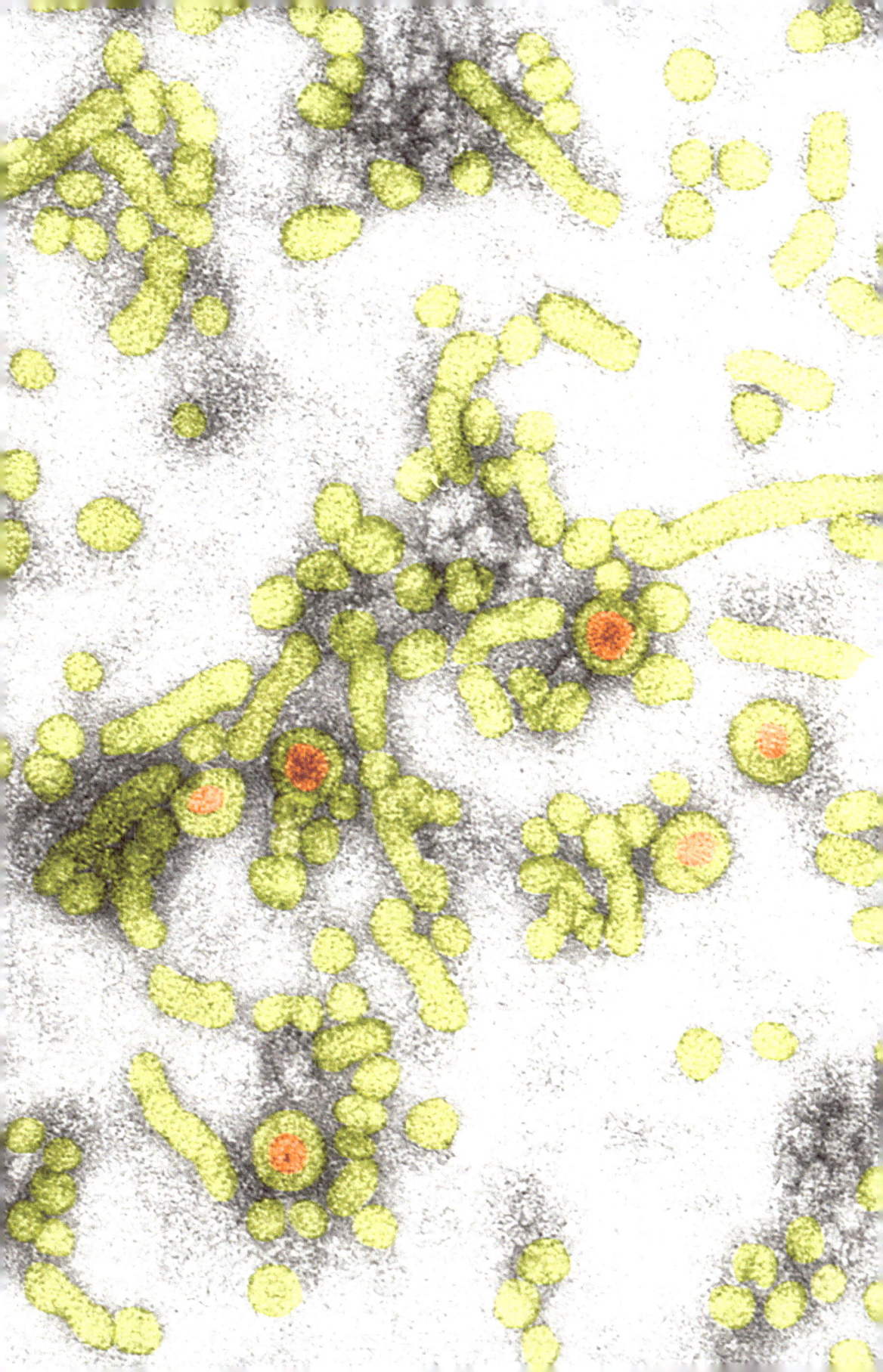

Panel 47. This image shows the subviral envelope particles of HBV (small spheres and filaments) constituting the early hepatitis B vaccine. This first-generation vaccine, developed by Philippe Maupas and his team in Tours, was made by the purification of the suviral envelope particles from the blood of HBV chronic carriers. This preparation shows that the virus was efficiently eliminated and that the vaccine only contained subviral particles, mainly spherical particles and a few filaments. In the mid-1980s, this vaccine was gradually replaced by second-generation vaccines, produced by genetic engineering. Indeed, the bioproduction of the HBV envelope protein by cultured cells allows the generation of excess envelope particles that can be easily purified and serve the same role as a vaccine. This "recombinant" vaccine, obtained by genetic engineering, was easier to produce and has therefore replaced the "plasma" vaccine. However, the first-generation vaccine had the merit of establishing the proof of concept of a safe and extremely effective vaccine.

The Hepatitis B vaccine

Over the past few decades, the hepatitis B vaccine has contributed to reducing the percentage of HBV chronic carriers in the world, at least in countries where vaccination campaigns are effectively conducted. HBV is mainly transmitted through blood and sexual secretions, as well as from mother to child. In a country like China, where newborn vaccination campaigns have been made mandatory, the number of HBV chronic carriers in the population has significantly decreased. The hepatitis B vaccine is often presented as the first "anti-cancer" vaccine, as it has significantly reduced the incidence of liver cancer in populations where the prevalence of HBV was high before vaccination campaigns. Unfortunately, this vaccine is not yet used systematically everywhere in the world. Overall, it is estimated that 20% of human cancers are due to infections by viruses (all types of viruses combined), and HBV still contributes significantly to this high percentage.

© The University of Tours 2025
P. Roingeard, *Journey to the Viral World: Electron Micrographs of Viruses,*
https://doi.org/10.1007/978-3-031-77995-4_45

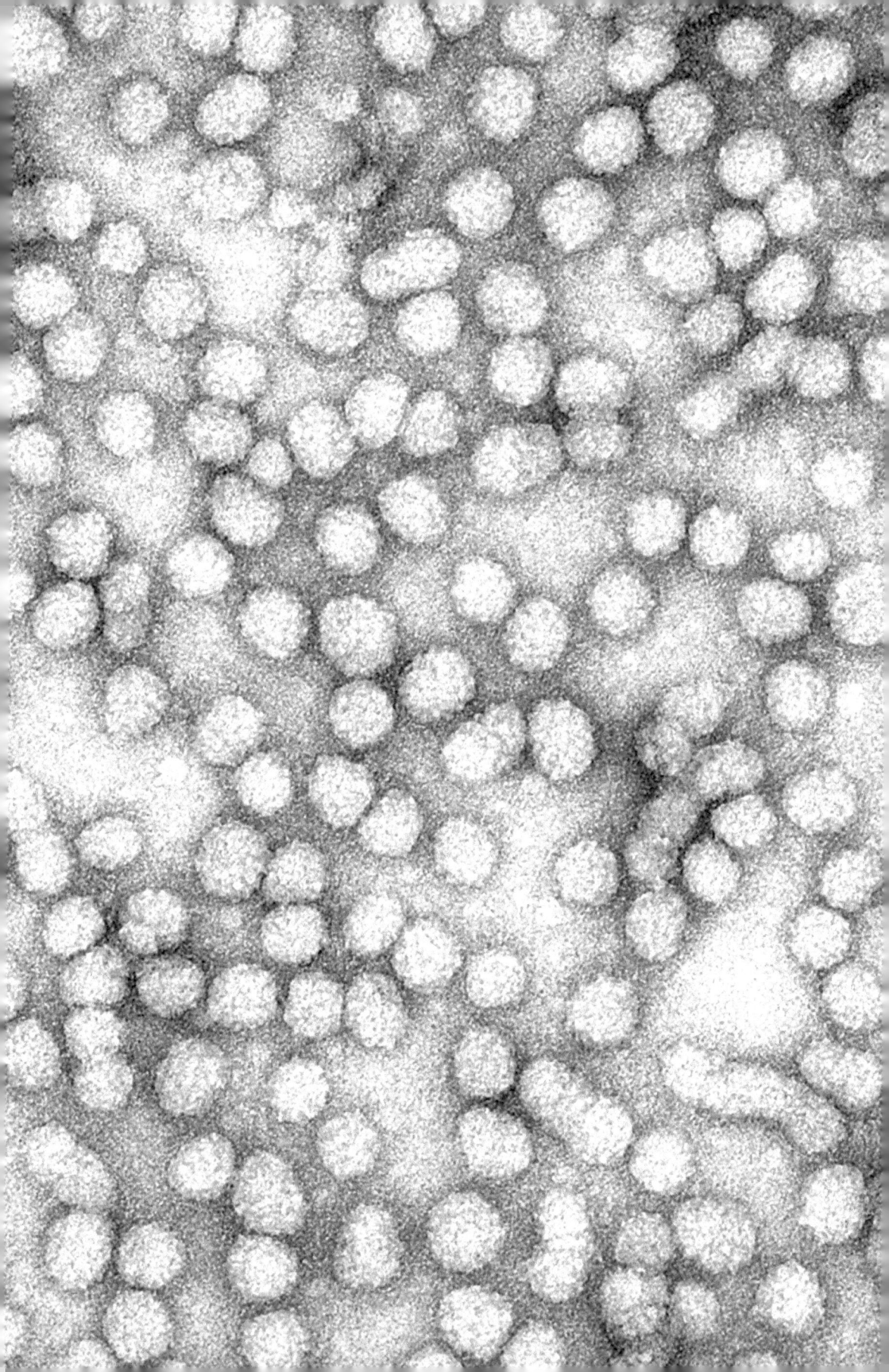

Panel 48. The formation of HBV excess envelope particles or "subviral" particles is quite astonishing. These images were made from cultured cells producing HBV vaccine particles. The two images at the bottom show cross-sections of these cells. The subviral particles are in the form of filaments compacted in cellular vesicles (image on top), which transport these filaments to a dilated cellular compartment where the filaments are released from this compaction (image on the bottom). The opposite image shows these filaments in solution, by negative staining. These photographs suggest that the HBV subviral particles are originally made up of filaments that are gradually converted into spherical particles. This probably explains why in the blood of chronically infected subjects contains both spherical and filamentous forms of subviral particles.

These images illustrate basic research that aimed to better understand the mechanisms of formation of these HBV subviral particles. This fundamental research has been very useful for much more applied research aimed at establishing a vaccine concept that can protect against both Hepatitis B and Hepatitis C. The principle of this vaccine was to produce by genetic engineering "chimeric" envelope proteins of the two viruses (HBV and HCV) capable of also forming vaccine particles. These particles therefore contain both the HBV and HCV envelope proteins and are the basis of a vaccine that can induce immune protection against these two viruses.

Excess envelope particles associated with the Hepatitis B virus

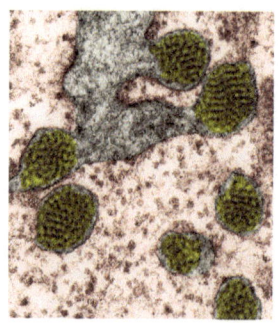

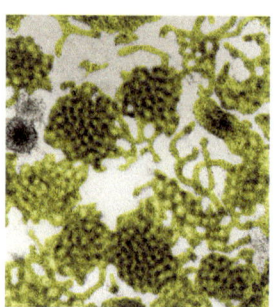

P. Roingeard, *Journey to the Viral World: Electron Micrographs of Viruses*, https://doi.org/10.1007/978-3-031-77995-4_46

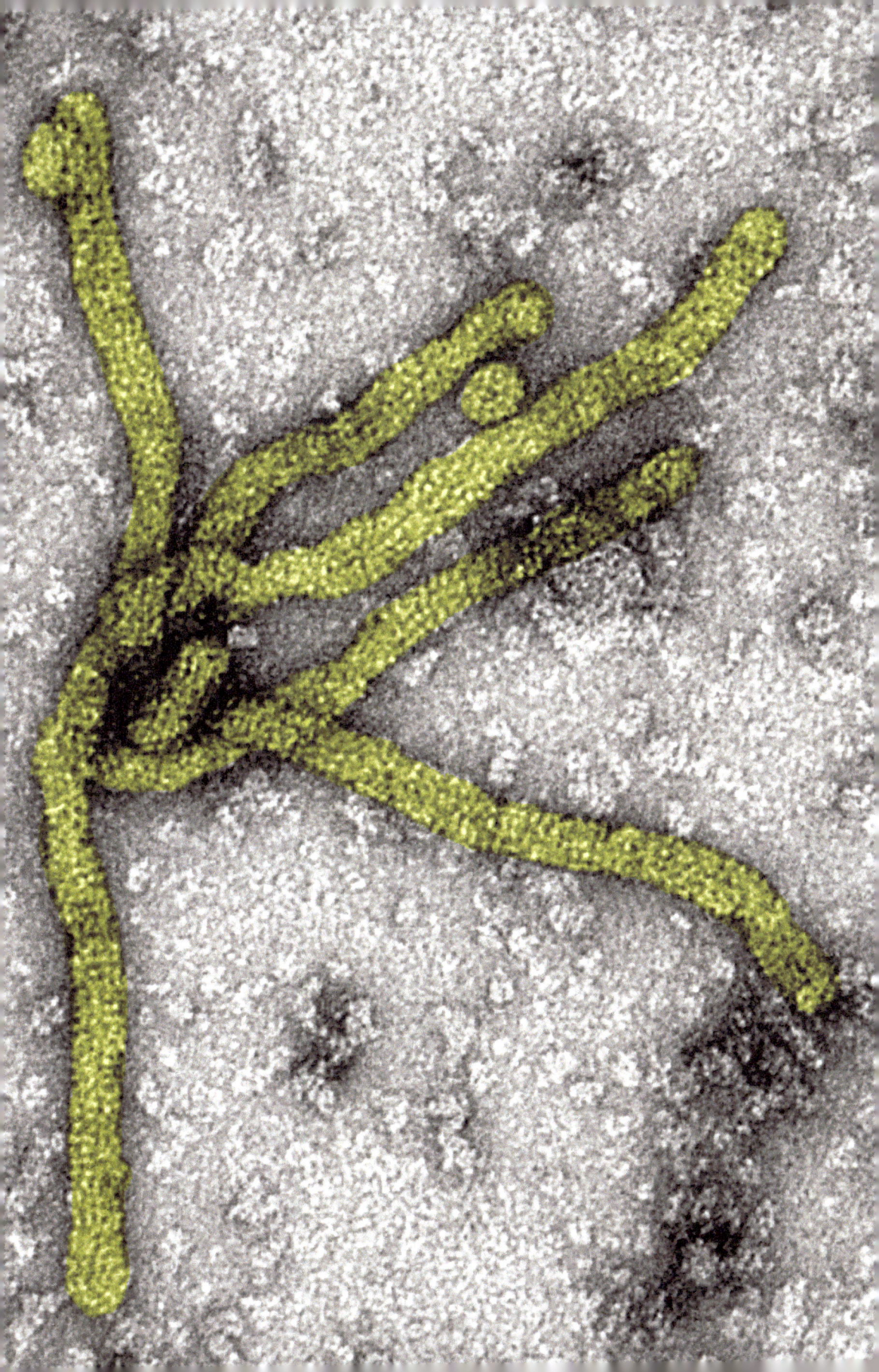

Panel 49. The Hepatitis C virus (HCV) is a very unique virus, whose originality lies in its close relationship with lipids. Cells can store lipid reserves in the form of spherical organelles called "lipid droplets". Depending on the needs of the cell, lipid reserves can be used or on the contrary increase, which influences the number and the size of the lipid droplets in the cell. This is particularly the case in liver cells (hepatic cells) in which the number of lipid droplets can fluctuate depending on the context (for example, after a fat-rich meal, their number and size increases). In a very curious way, the HCV, which specifically infects hepatic cells, uses the lipid droplets of the infected cells as an assembly platform. This photograph shows a cross-section of a cell producing HCV. Two large lipid droplets are colored in yellow. They are surrounded by a dilated cellular compartment (in purple) in which the recently assembled viruses are present (the small dark particles inside this compartment). The cytoplasm of the cell is colored in light blue.

Hepatitis C virus and lipids

Research has shown that the virus associates with lipid droplets to "capture" certain cell compounds essential to its formation, notably apolipoproteins. These apolipoproteins are essential not only to the formation of the virus, but also to its secretion out of the infected cell. Apolipoproteins are indeed naturally secreted by liver cells. They then enter the bloodstream to ensure the transportation of lipids such as cholesterol throughout the body. HCV uses apolipoproteins to take advantage of this secretory flow to exit the cell and then circulate in the blood of infected individuals. This is yet another example of the ability of viruses to exploit cellular mechanisms to their advantage.

© The University of Tours 2025
P. Roingeard, *Journey to the Viral World: Electron Micrographs of Viruses*,
https://doi.org/10.1007/978-3-031-77995-4_47

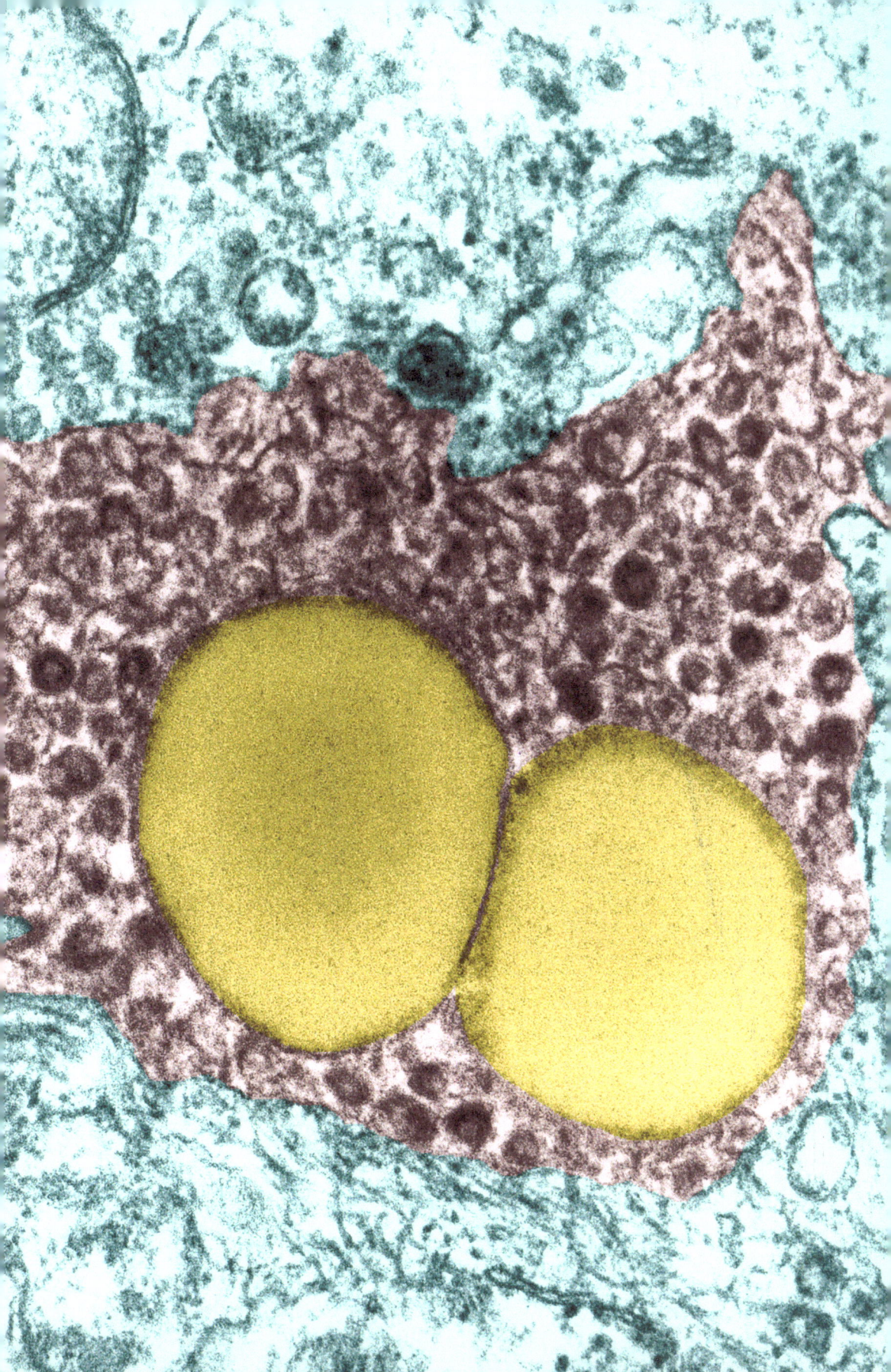

Panel 50. The hijacking of lipid droplets by the Hepatitis C virus to ensure its formation and exit mechanisms from the cell is not without clinical consequence for people chronically infected with this virus. HCV is indeed capable of inducing an increase in the number and size of the lipid droplets contained in infected cells. This accumulation of numerous and sometimes giant lipid droplets in the liver cells leads to a pathology called liver steatosis. This strong accumulation of lipids inside the liver cells induces a liver dysfunction which contributes to the acceleration of the disease in patients chronically infected with HCV. This image shows a section of a liver sample from a patient suffering from chronic hepatitis C with steatosis. This liver cell contains several giant lipid droplets (in yellow-green) that invade the entire cytoplasm. These lipid droplets are so large that they take up all the space in the cell, pushing out the nucleus (in purple), which ends up taking the shape of the droplets.

Hepatitis C virus and liver steatosis

© The University of Tours 2025
P. Roingeard, *Journey to the Viral World: Electron Micrographs of Viruses*,
https://doi.org/10.1007/978-3-031-77995-4_48

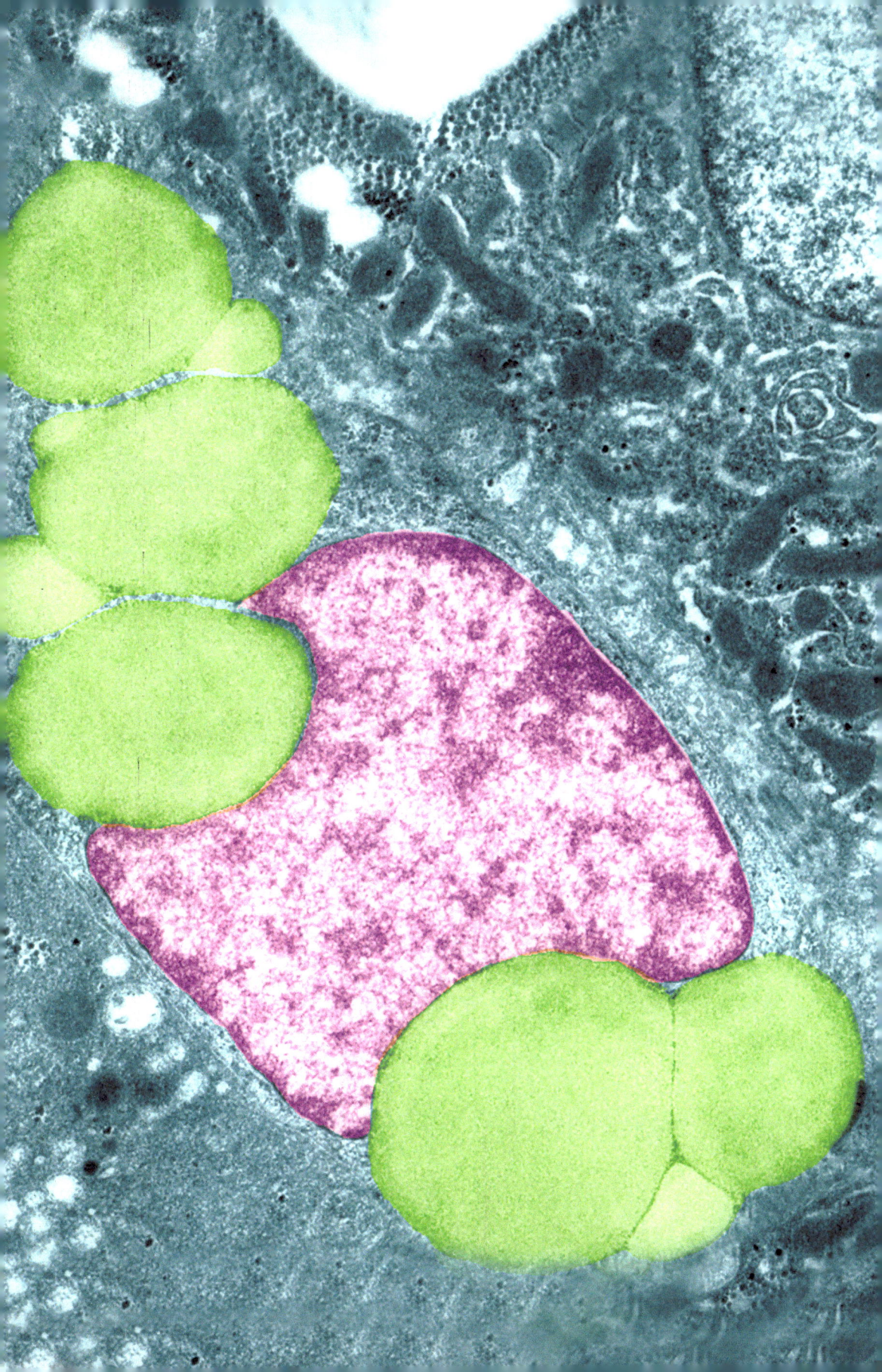

Panel 51. It took more than 25 years between the discovery of HCV (in the early 1990s) and its visualization by electron microscopy by negative staining, by young researchers from my team, in 2016. Several reasons explain this very long wait. Initially, this virus could only be discovered thanks to molecular biology techniques. Its **RNA genome** was therefore known well before its proteins and biological characteristics. It was not until 2005 that the first cellular systems were available to cultivate the virus in the laboratory. Moreover, this virus proved to be fragile with a tendency to degrade when attempts were made to purify it. Finally, its curious association with apolipoproteins meant that it was long confused with lipoproteins circulating in the blood.

This image shows a HCV particle (top right), accompanied by three other particles. It has been colorized to show the difference between (i) the virus with its **capsid** (in orange) and its envelope (in yellow); (ii) and simple circulating lipoproteins (in yellow). This observation was made by capturing these particles with specific antibodies (and therefore without a prior purification step that weakens the virus) in the blood of a patient chronically infected with HCV. HCV resembles a circulating lipoprotein, but also contains a **capsid** inside. This **capsid** is not conventional as it is poorly structured, unlike most viruses. Its envelope is also atypical, as it is very rich in lipids and gives this "bloated" appearance. The size of the virus can vary from one patient to another, depending on the presence of lipids in their blood in greater or lesser quantities.

Hepatitis C virus morphology

© The University of Tours 2025
P. Roingeard, *Journey to the Viral World: Electron Micrographs of Viruses,*
https://doi.org/10.1007/978-3-031-77995-4_49

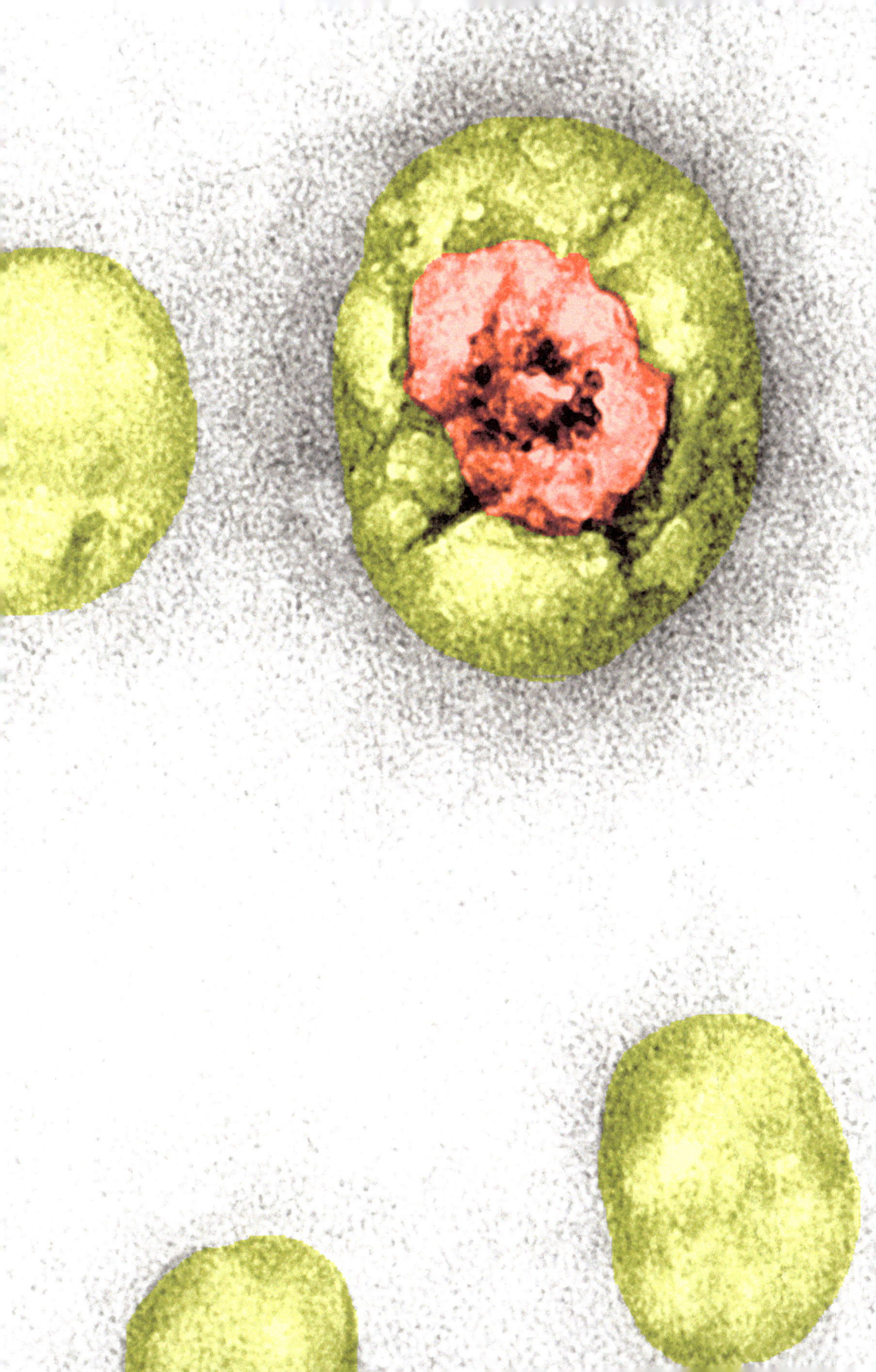

Panel 52. The West Nile virus is a small **RNA** virus, enveloped, 1/50000 th of a millimeter in diameter, belonging to the Flavivirus family, which also includes the yellow fever virus, the dengue virus (see panel 54), or the Zika virus (see panel 55). This photograph shows a small group of West Nile viruses (in green), present in a cellular compartment at the center of this cell section. The virus gets its name from the "West Nile" district, a region in Uganda where the virus was first isolated in the 1930s.

This virus primarily infects wild and domestic birds, in which it replicates with great efficiency. But it also infects mammals: horses, dogs, cats, domestic rabbits, and of course humans, transmitted by bites from mosquitoes of the *Culex* genus.

In humans, most infections do not cause symptoms, but about 20% progress to a flu-like syndrome characterized by muscle pain, headaches, and respiratory problems. About 1% of infections can cause **encephalopathy**. Medical historians believe that the conditions of Alexander the Great's death in Babylon in 323 before Christ had all the

The West Nile Virus

characteristics of an **encephalopathy** induced by a West Nile virus infection, showing that this virus has likely been infecting humans for millennia.

The virus is mainly present in the tropical areas of Africa, but it can travel to the temperate zones of Europe, Asia, and the American continent through migratory birds. It can then spread through bites from local mosquitoes to other animals and humans. However, it does not transmit from human to human through mosquito bites. Southern France has experienced several epidemics of West Nile virus infection in horses, with a few cases of secondary transmission to humans. Mini-epidemics have also occurred in Eastern Europe, Israel, and North America in recent years. Global warming raises fears that this virus may in the future more easily spread in populations, due to a change in the geographical range of the tiger mosquito, which is a particularly effective vector of the West Nile virus, as well as other Flaviviruses.

P. Roingeard, *Journey to the Viral World: Electron Micrographs of Viruses*, https://doi.org/10.1007/978-3-031-77995-4_50

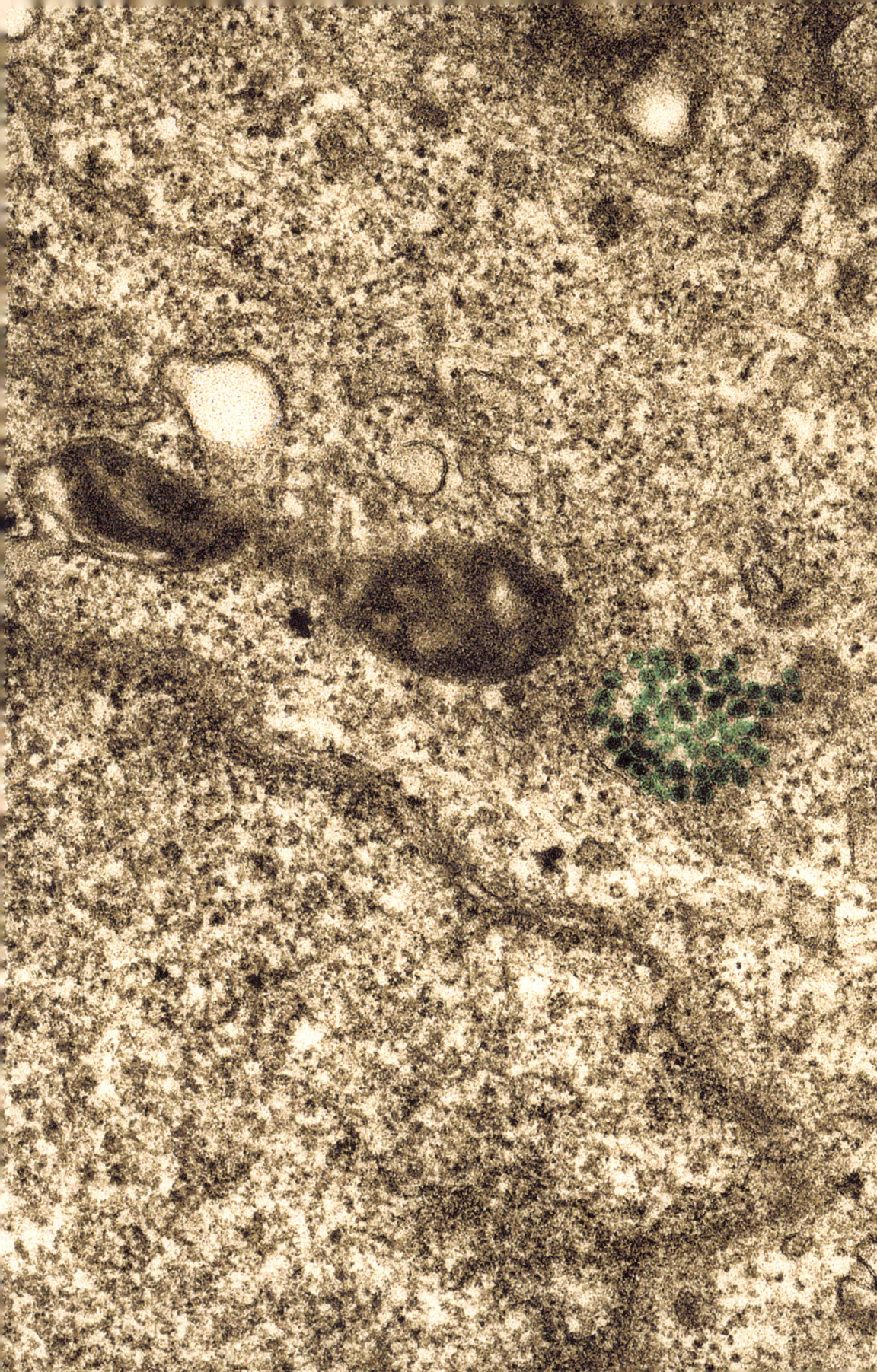

Panel 53. Like all Flaviviruses, the West Nile virus induces profound rearrangements of the infected cell membranes, as observed in these images showing small circular structures delimited by a membrane (in pink). These small compartments are real "factories" ded-

icated to large-scale replication of the **viral genome**. Their membranes are of cellular origin, but these compartments do not normally exist in the cell. The virus hijacks the machinery of cellular membrane synthesis for its benefit to create these small structures. They constitute an environment for optimal replication of the

Membrane rearrangements and

West Nile Virus

viral RNA, by grouping inside all the essential constituents for this activity. The membrane that delimits these "factories" of **viral genome** production serves to isolate them from the rest of the cell and thus escape the internal defense mechanisms of the cell.

These photographs shows also areas of the cell that contains both these "factories" for **genome** replication (in pink) and West Nile viruses (in green), just formed in other internal cellular compartments. These images of "viral factories" show once again how much viruses can hijack cellular machinery for their benefit.

© The University of Tours 2025
P. Roingeard, *Journey to the Viral World: Electron Micrographs of Viruses*,
https://doi.org/10.1007/978-3-031-77995-4_51

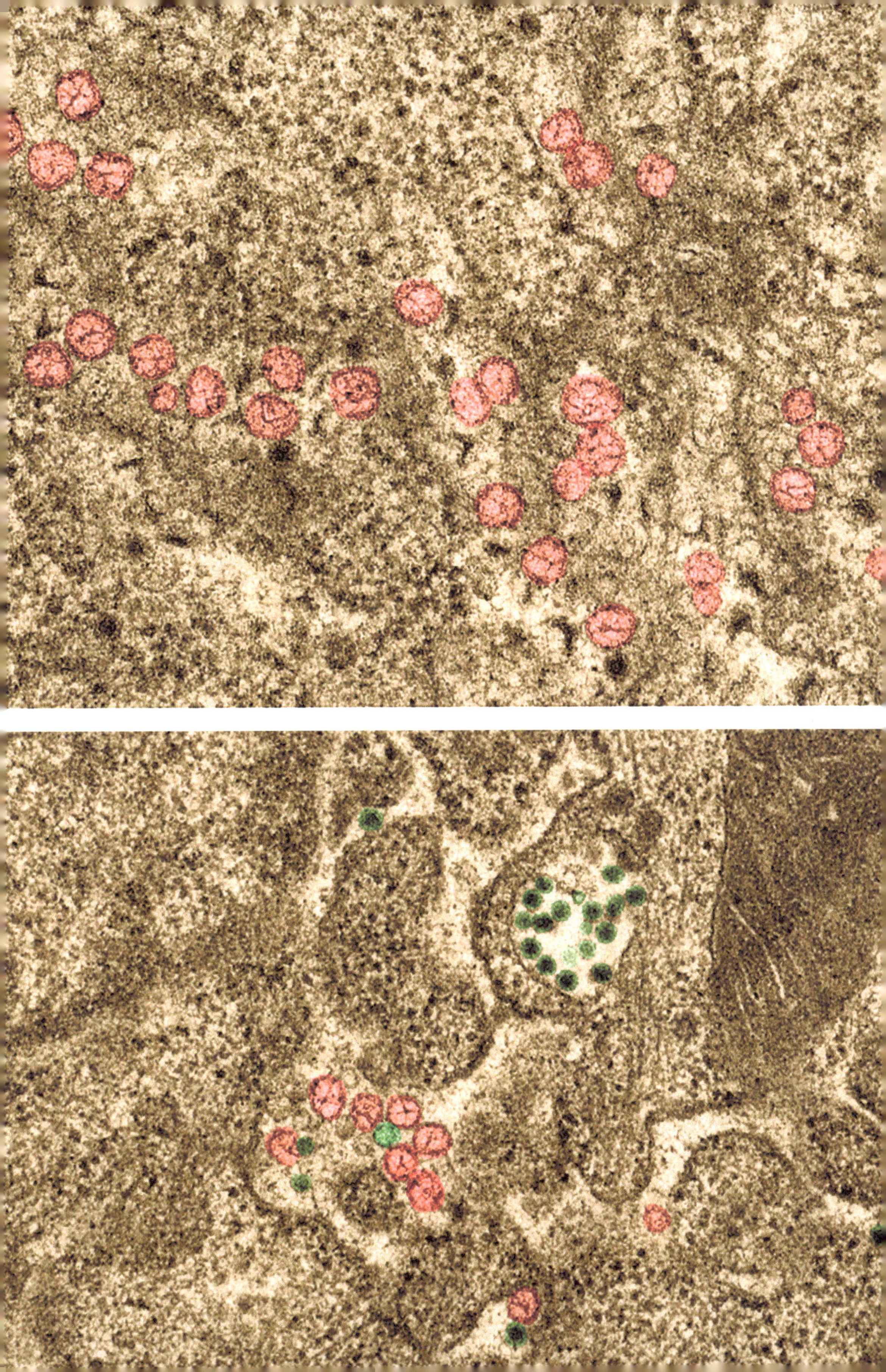

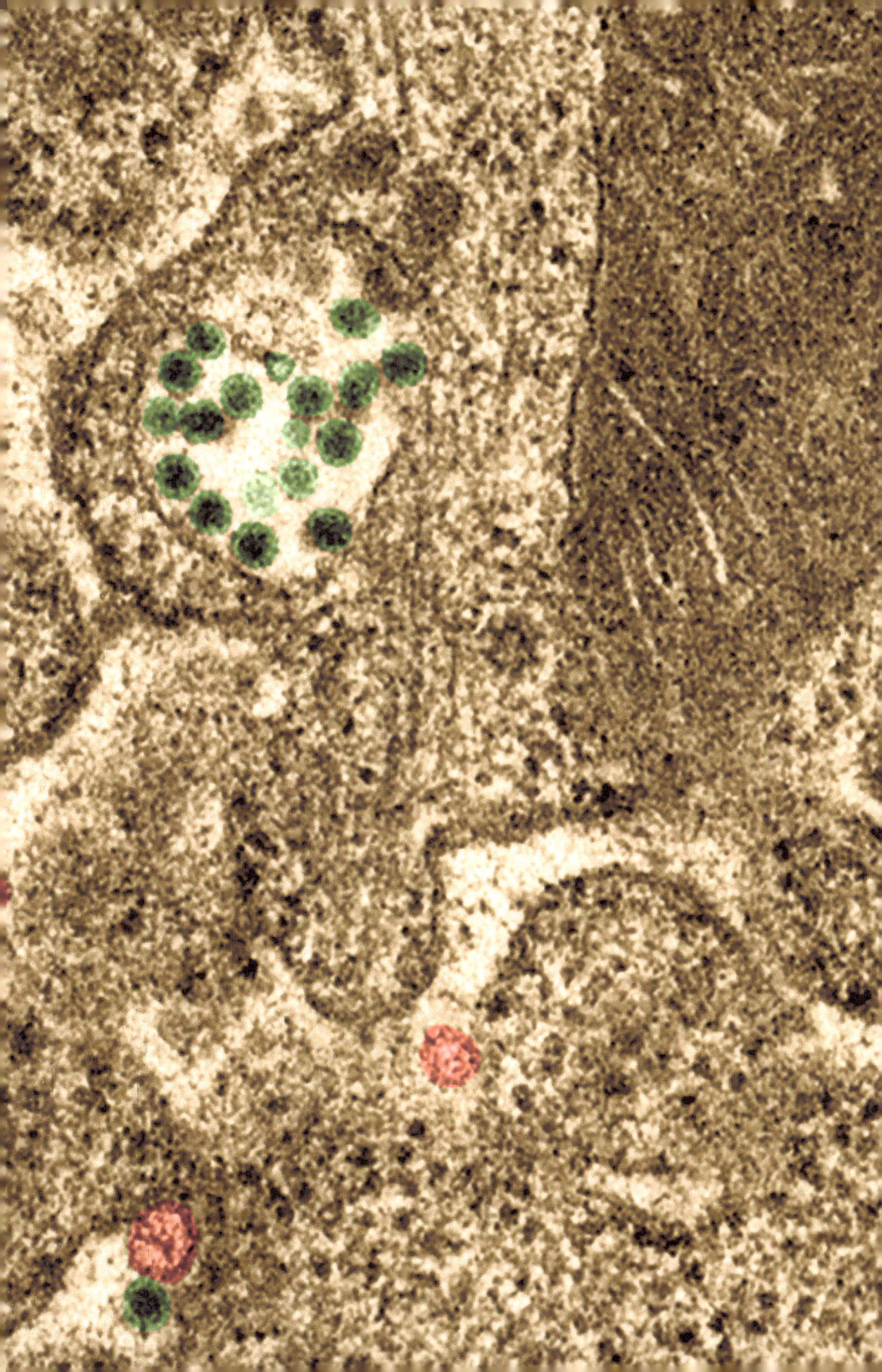

Panel 54. The dengue virus represents another member of the Flavivirus family, also transmitted by mosquito bites. These images show sections of cells infected by this virus. The photograph at the top shows multiple membrane rearrangements that are the site of the replication of the **viral genome**. They are slightly different morphologically from those induced by the West Nile virus, but their role is identical. The photograph at the bottom shows an area of this cell that contains membrane rearrangements and viral particles, at high magnification. The morphology of this virus is comparable to that of the West Nile virus. However, infection by this virus has much more significant consequences than in the case of the West Nile virus. The pathology is not only more severe, but the dengue virus is efficiently transmitted from human to human, with the human species constituting the reservoir of the virus. Moreover, there are 4 subtypes of this virus, and while infection by one subtype can induce protective immunity against the same subtype, it does not prevent reinfection by another subtype.

The dengue Virus

The WHO estimates that, every year, nearly 400 million human beings are infected by this virus. About 100 million develop clinical signs of this infection, with high fevers, headaches, muscle pains. The name "dengue" proba-bly comes from the Spanish word denguero (stilted), relat-ing to the stiff gait of those suffering from muscle pain due to dengue virus infection. Among these individuals with clinical signs of infection, it is estimated that half a million develop a very severe form of the disease, characterized by hemorrhagic fevers. The estimated number of deaths due to these severe forms of dengue is about 15,000 per year worldwide.

The main vector of the virus is the mosquito *Aedes aegypti*, of African origin and present in the tropical and subtropical regions of the planet. This one is particularly effective in transmitting this virus. The tiger mosquito (*Aedes albopictus*), of Asian origin, is a little less effective in transmitting it, but it has the ability to adapt to more temperate regions, especially if their temperature rises slighlty due to global warming. This observation is certainly worrying for the future, showing the risk of seeing dengue spread more widely to the more temperate regions of the globe.

© The University of Tours 2025
P. Roingeard, *Journey to the Viral World: Electron Micrographs of Viruses,*
https://doi.org/10.1007/978-3-031-77995-4_52

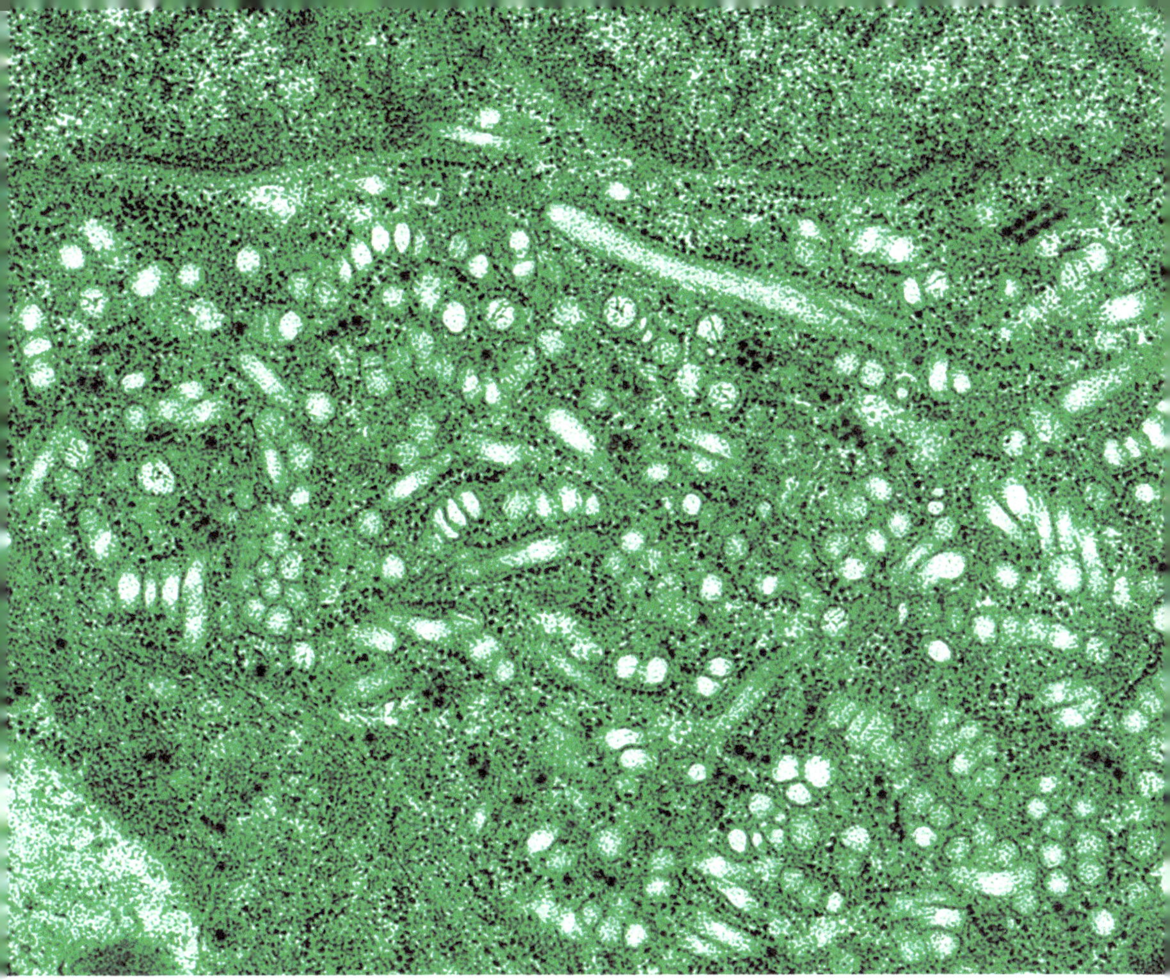

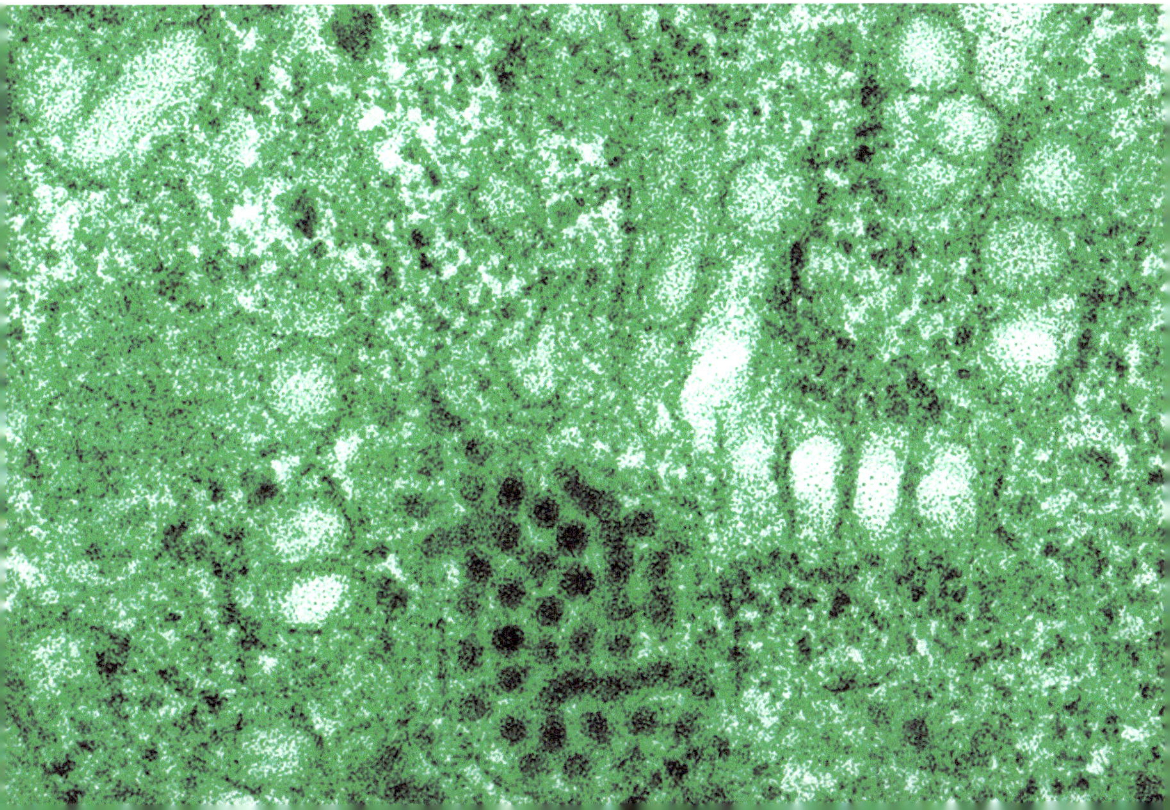

Panel 55. The Zika virus represents another member of the Flavivirus family, whose morphology and modes of multiplication are very similar to the West Nile virus (see panels 52 and 53) and dengue virus (see panel 54). Like these two other viruses, it is classified as an **arbovirus** (for **ar**thropod **bo**rne **virus**, meaning a virus that is transmitted by biting arthropods, mosquitoes or ticks), but it has the particularity of being able to be transmitted by many species of mosquitoes of the genus *Aedes*. Like the dengue virus, it is transmitted from human to human and therefore spreads in the human population through mosquito bites. Moreover, it is the only arbovirus for which a possibility of sexual transmission has been demonstrated. This image of a cross-section of an infected cell shows viral particles very similar to those

The Zika Virus

of the West Nile and dengue viruses. In the colored inset, a photograph of an isolated viral particle observed by negative staining shows this very compact enveloped virus, about 1/50000 th of a millimeter.

Its name come from the Zika forest, on the edge of Lake Victoria in Uganda, where it was first isolated in a monkey during the 1930s. Its presence in humans was only demonstrated 20 years later. It was long confined to the tropical regions of Africa and Asia, before emerging in the form of epidemics, since 2007, in Oceania, in French Polynesia, New Caledonia and especially the one that occurred in South America and Central America in 2015.

As with the other arboviruses of the Flavivirus family, infection with the Zika virus often gives very few symptoms, but can cause fevers, headaches, joint pain. In addition, infection with the Zika virus is associated with two types of serious complications: microcephaly (smaller than normal head) in newborns of infected mothers; on the other hand, Guillain-Barré syndrome, a neurological disease associated with motor disorders. These complications were not known before the South American epidemic of 2015, and were undoubtedly revealed by the scale of individuals affected by this infection during this epidemic (nearly 2 million).

© The University of Tours 2025
P. Roingeard, *Journey to the Viral World: Electron Micrographs of Viruses*,
https://doi.org/10.1007/978-3-031-77995-4_53

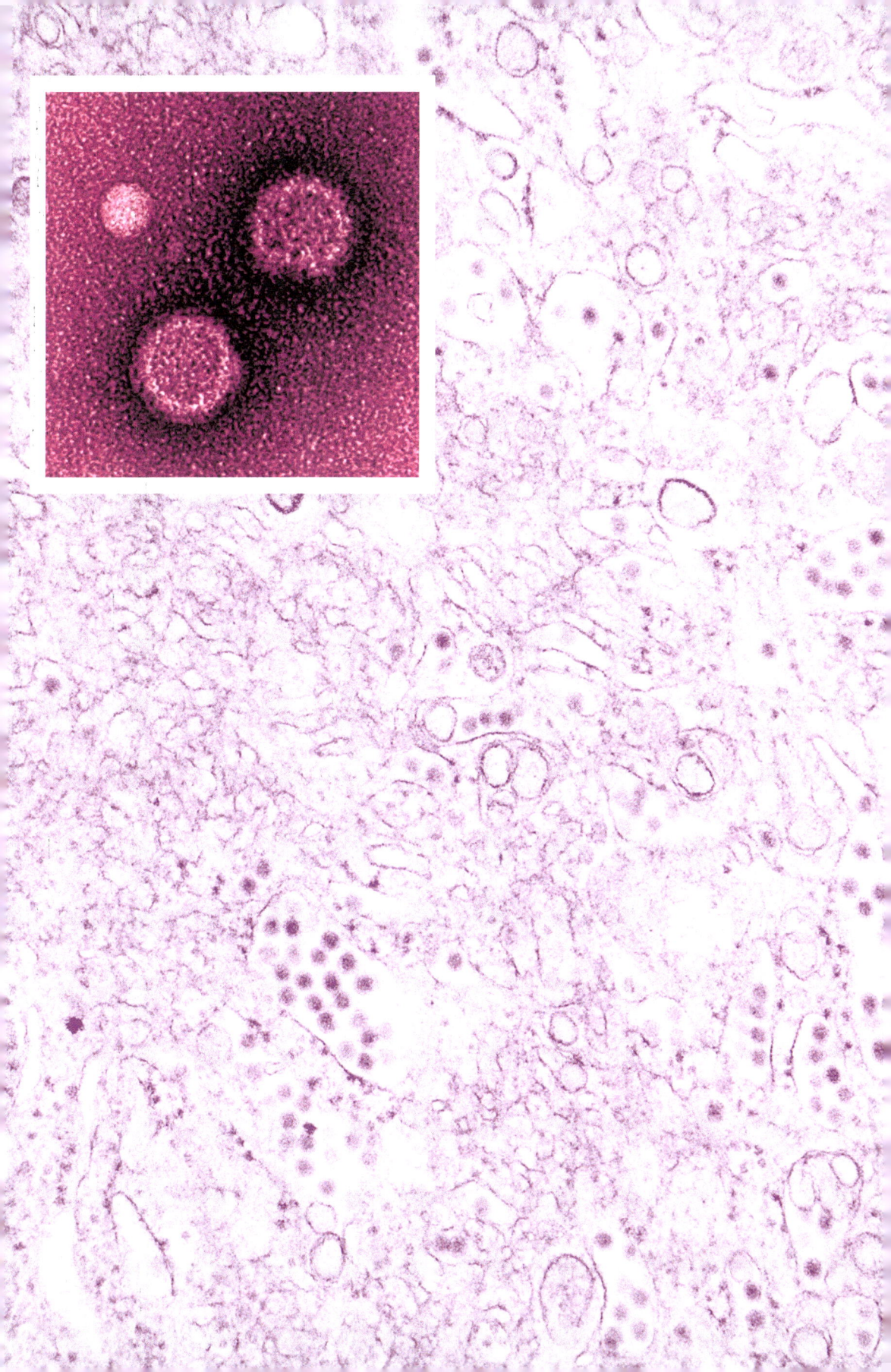

Panel 56. The Zika virus is an example of a virus that induces a very marked cytopathogenic effect, i. e. structural changes in a host cell resulting from the viral infection. As with other viruses, viral replication mobilizes most of the cell's biochemical resources, which are profoundly affected, and the cells exhibit clear morphological alterations.

Cytopathogenic effect
of the Zika virus

This cross-section of a cell infected with the Zika virus shows giant intracelluar vacuoles induced by the virus. These large and empty vacuoles correspond to cellular compartments that swell abnormally, eventually leading to an "implosive" cell death. Although enveloped viruses can exit a cell without causing it to burst, it is certain that this type of mechanism contributes to many Zika viruses exiting the cell at the same time, facilitating their spread throughout the body.

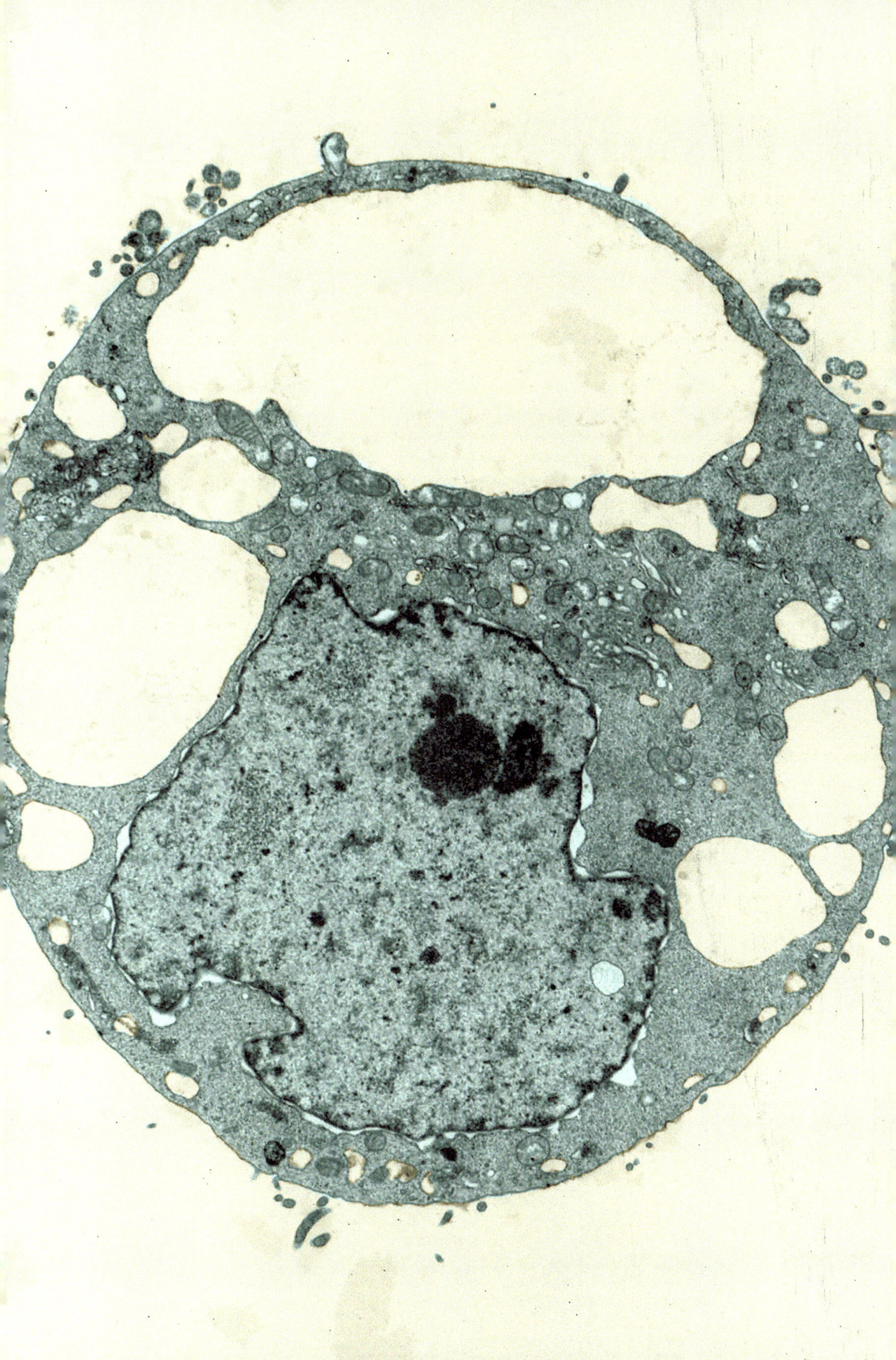

Panel 57. Chikungunya virus is an **arbovirus** whose morphology is reminiscent of the arbo-viruses in the Flavivirus family (West Nile, Dengue, Zika, see panels 52 to 55), but it belongs to another family, the Togaviruses (which also includes the Rubella virus, despite its very different appearance; see panel 43). It has many other similarities with dengue, apart from its morphology: it is transmitted by the mosquitoes *Aedes aegypti* and *albopictus,* from human to human, and it induces fevers, headaches and muscle and joint pains. However, its main characteristic is to induce particularly strong and therefore very disabling joint pains, which can persist for months, or even years. The word "chikungunya" comes from a local language spoken in East Africa, meaning "that which bends up". This is indeed the "disease of the bent man", due to the joint pains induced by the infection, which literally bend the infected subjects.

The chikungunya Virus

In this photograph of an infected cell, numerous viral particles (in green) cover the entire surface of the cell (in yellow). This presence of the virus on the cell surface constitutes a difference with the West Nile, Zika and dengue viruses, which form in cellular compartments and then leave the cell following the secretory flow. Chikungunya viruses form by budding on the cell surface.

Despite these symptoms, infection with this virus evolves in the majority of cases favorably. However, recent epidemics affecting many people revealed neurological complications in elderly people or newborns of infected mothers. Indeed, while the virus has been known since the 1950s and seemed confined to regions of East Africa, epidemic waves have occurred in the islands of the Indian Ocean (notably on Reunion Island in 2005, where a third of the population had been affected) and Central America (Guadeloupe and Martinique in 2014). The tiger mosquito (*Aedes albopictus*) tending to gain temperate regions, it will probably be possible to observe mini-epidemics occurring in Europe in the coming years. Chikungunya virus infections have already been reported in Italy and in the south of France.

© The University of Tours 2025
P. Roingeard, *Journey to the Viral World: Electron Micrographs of Viruses,*
https://doi.org/10.1007/978-3-031-77995-4_54

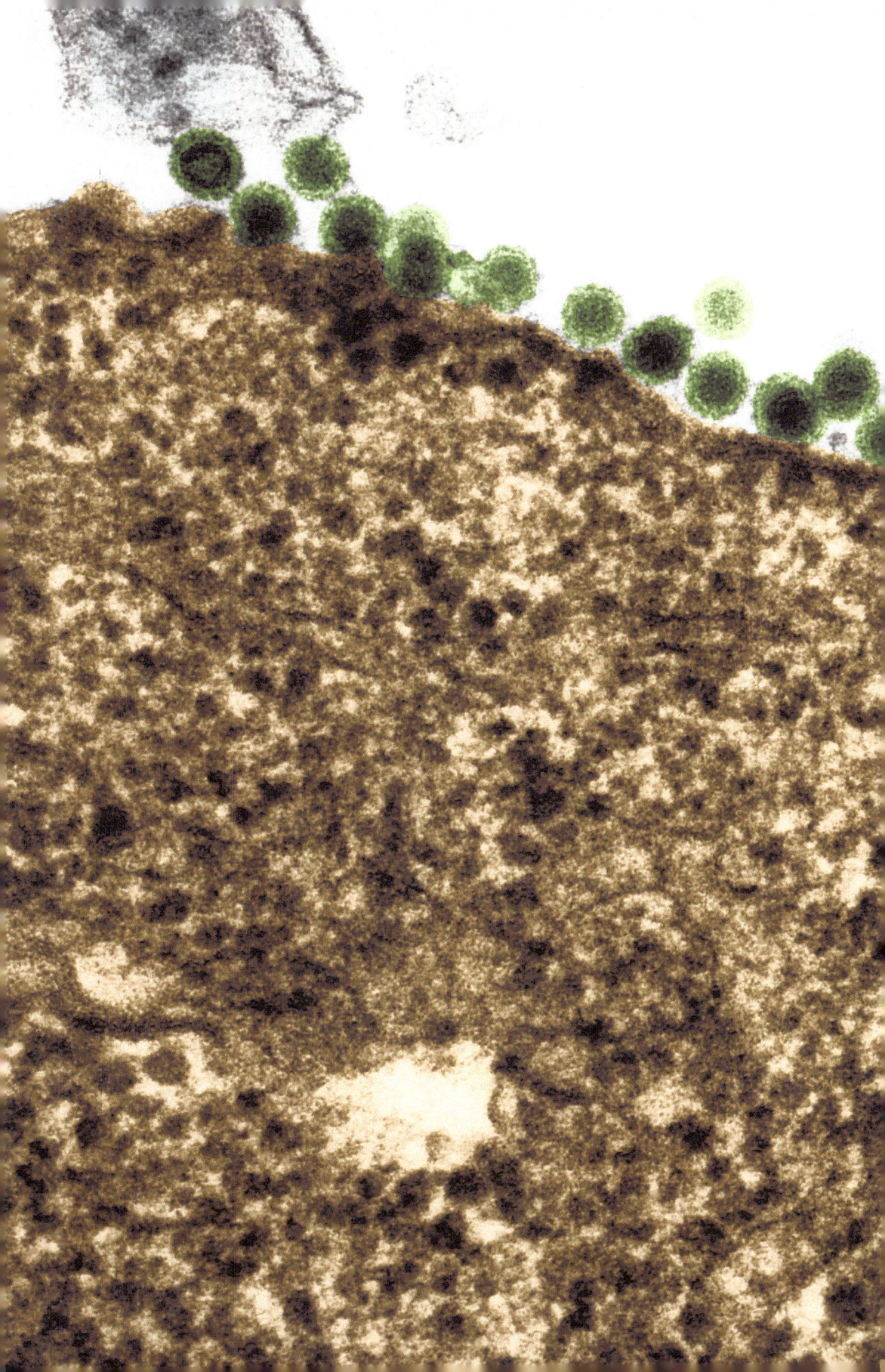

Panel 58. The structural modifications of a cell infected which the chikungunya virus are very original in the world of viruses, as shown by these photographs of different sections of infected cells.

The photograph opposite shows replication vesicles on the surface of an infected cell (the group of pink vesicles; the two isolated particles in green are viral particles that have just left the cell). These small vesicles are "factories" for viral replication, with similar functions as those seen for the West Nile virus (panel 53) or the dengue virus (panel 54). But for the chikungunya virus, they are located at the surface of the cell and remain anchored to its plasma membrane. Again, the virus uses this strategy to hijack the cellular machinery for its benefit, exploit the cell's membranes, and escape the cellular defense mechanisms. The following photographs show viral **capsids** (in green) before leaving the cell. They are anchored to the membrane of intracellular compartments to form rosette-like structures. The cellular membranes involved in these structures belong to the Golgi apparatus, an organelle involved in the cell's secretion mechanisms. In physiological conditions, secretion vesicles generated from the Golgi apparatus fuse with the cell's plasma membrane

The formation of the

chikungunya Virus

to release synthesized compounds into the extracellular environment. These "Golgian" vesicles also serve to deliver the future constituents of the plasma membrane. In the case of an infected cell, the Golgi membranes on which the viral **capsids** are anchored also contain the viral envelope proteins. These Golgi membranes containing the viral proteins are brought to the cell's plasma membrane so that the virus can form, by budding of the viral capsid through these membranes containing the viral envelope proteins. The virus is then released outside the cell. These mechanisms once again illustrate the ability of viruses to hijack the cellular machinery in multiple ways.

© The University of Tours 2025
P. Roingeard, *Journey to the Viral World: Electron Micrographs of Viruses*,
https://doi.org/10.1007/978-3-031-77995-4_55

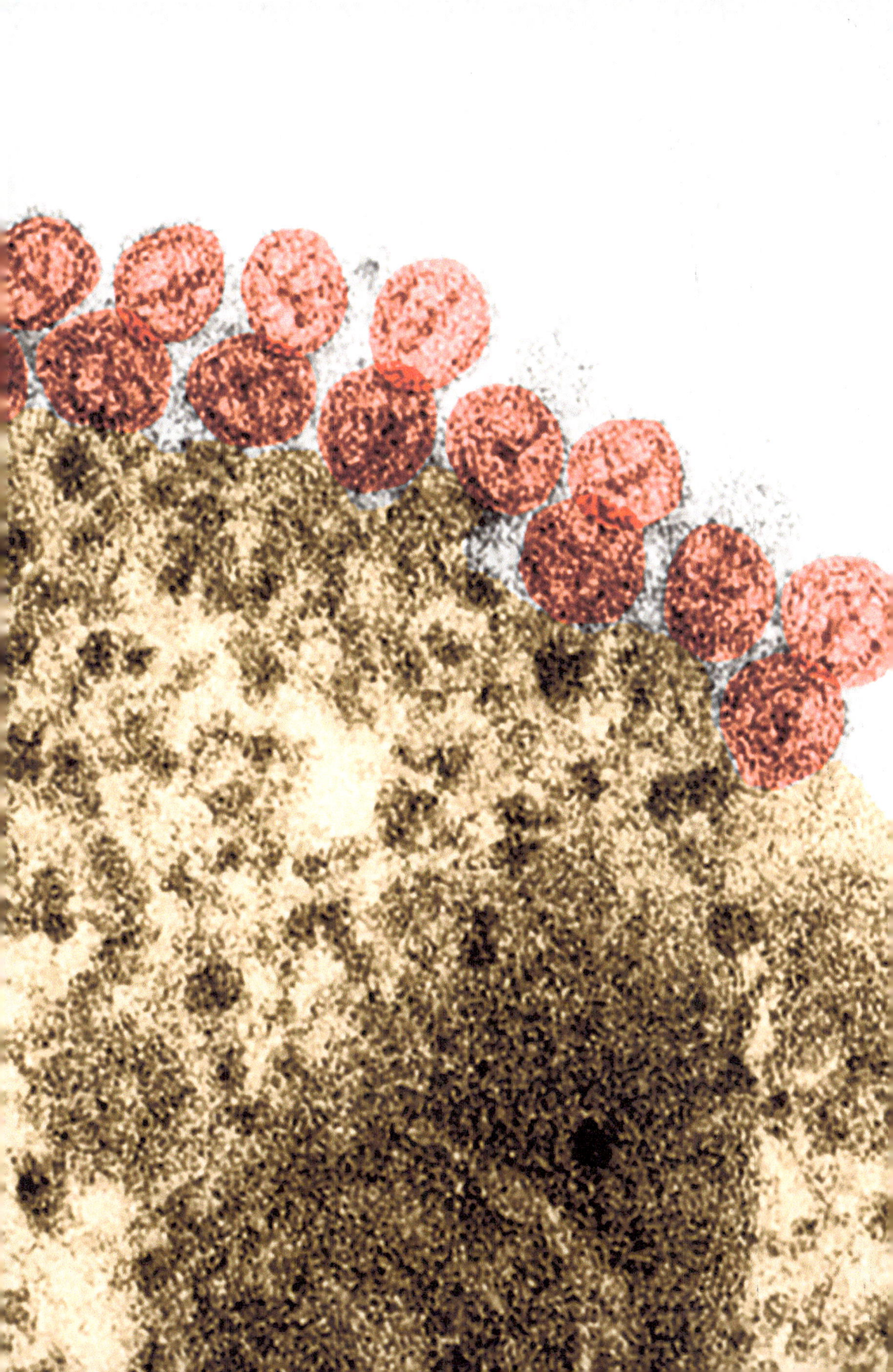

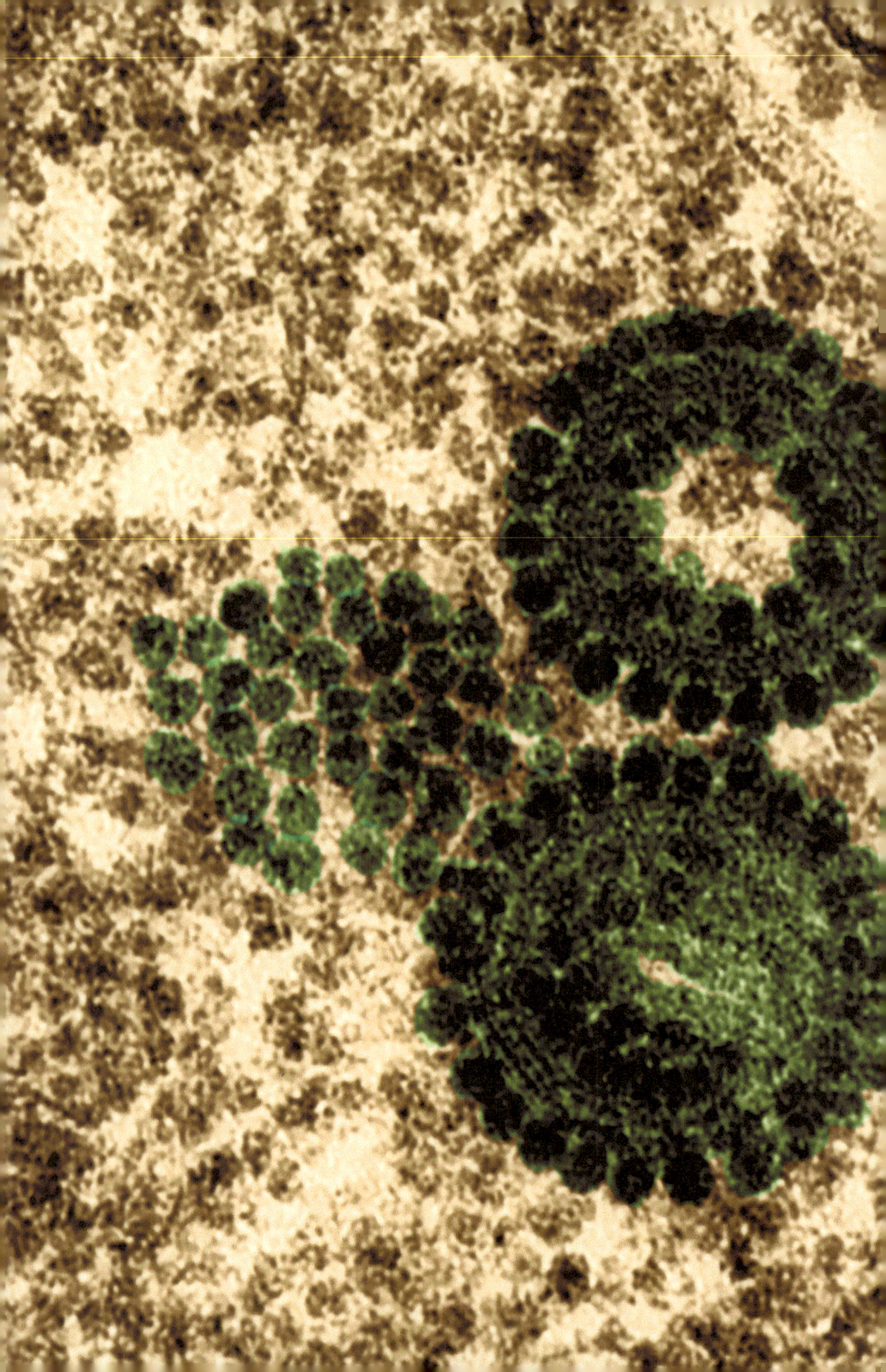

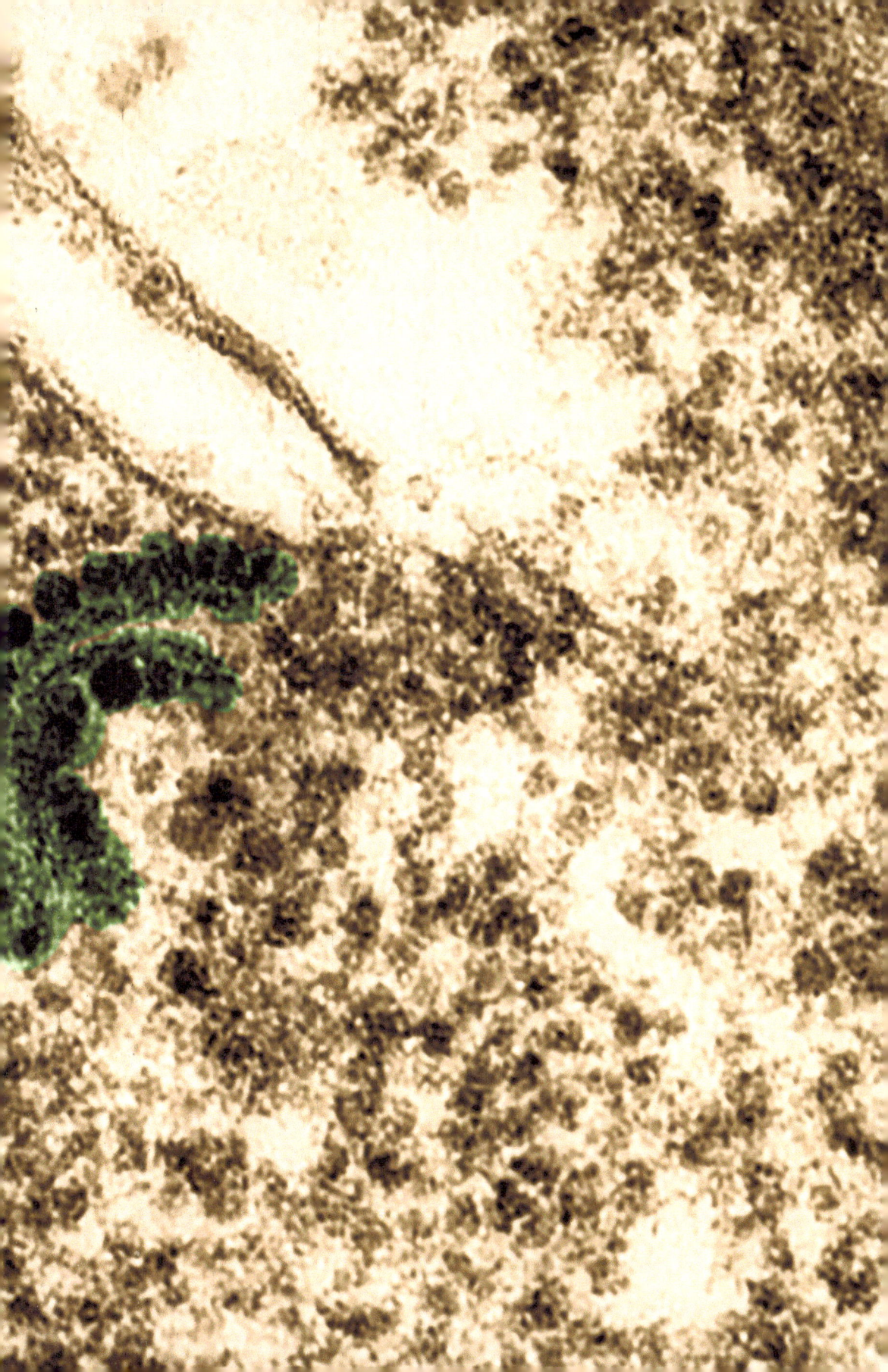

Panel 59. Hantaviruses are enveloped RNA viruses named after the Hantaan River, which separates North and South Korea. These viruses were indeed responsible for hemorrhagic fevers associated with renal diseases in American soldiers during the Korean War, between 1951 and 1954. However, it took more than 20 years before the virus was identified in small rodents in this region. Since then, several viruses from this family have been identified in small rodents or bats in Africa, Europe, or South and North America. The infected animals do not develop any disease, but the virus is present in their feces, urine, and saliva. In case of contact with these excretions, these viruses can be transmitted to

Hanta-viruses

humans. The consequences are very variable, some viruses inducing hemorrhagic fevers with renal syndrome (as in Korea), others pulmonary diseases (for those on the American continent). Others do not induce any disease.

This image shows a cell (in green) infected by the Tulavirus (in orange), a Hantavirus present in small rodents in Europe, but for which no associated disease is known. The virus presents a rather atypical shape with spherical particles about 1/10000th of a millimeter extended by filamentous structures of 1/50000th of a millimeter in diameter and of variable length.

© The University of Tours 2025
P. Roingeard, *Journey to the Viral World: Electron Micrographs of Viruses*,
https://doi.org/10.1007/978-3-031-77995-4_56

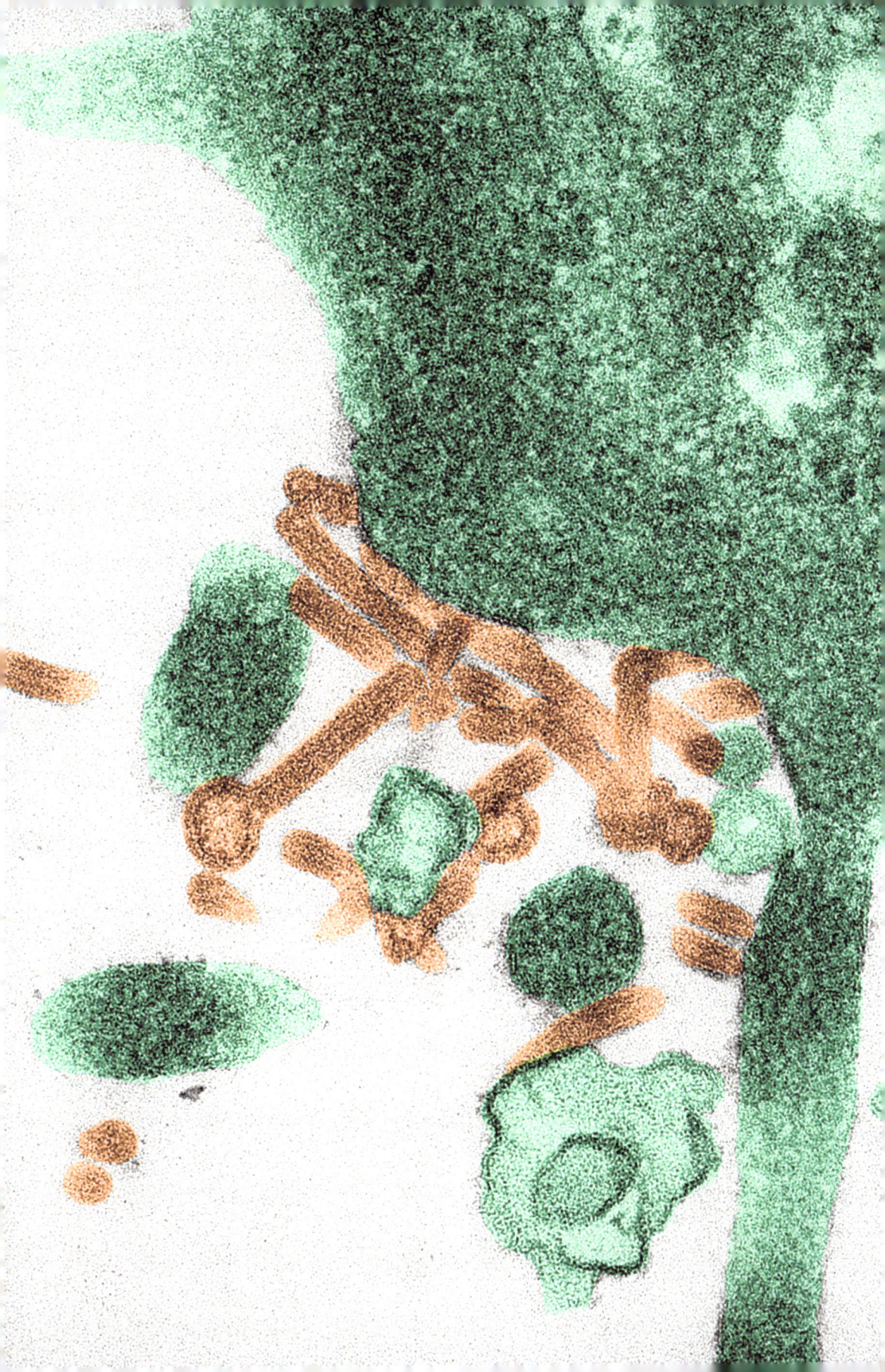

Panel 60. Coronaviruses are enveloped viruses, about $1/10000^{th}$ of a millimeter in diameter, with a **capsid** of **helical** symmetry surrounding a large **RNA genome**. They get their name from their appearance under electron microscopy, as they seem to be surrounded by a crown (corona meaning crown in Latin). This crown is due to particularly protruding envelope proteins exposed on the surface of these viruses. This large viral family includes viruses that infect the digestive and respiratory tracts of birds and various mammals, including humans. The coronaviruses that usually infect humans are relatively harmless, causing only non-serious colds. However, one of the major problems posed by this family of viruses in public health is that animal viruses have occasionally been able to contaminate humans and induce, on the occasion of this species barrier crossing, very severe pathologies. As these viruses can mutate rapidly, they have been able to adapt to human-to-human transmission and spread rapidly in populations during these episodes. Thus, an animal coronavirus was the cause of the SARS (Severe Acute Respiratory Syndrome) epidemic that hit China in 2003, contaminating more than 8000 people and causing nearly 700 deaths. The virus was found to be transmitted from human to human through the air, probably by contaminated saliva droplets. Air travel facilitated its global spread, with SARS cases concentrated in airport hubs or in areas with high population densities. In 2012, another animal coronavirus was the cause of a respira-

Corona-
viruses

tory syndrome also causing nearly 500 deaths, called MERS, for Middle East Respiratory Syndrome, as it originated in the Middle East. In both cases, the natural reservoir animals of these viruses appeared to be bats, the virus having been transmitted to humans via intermediate animals in contact with humans (the civet, a kind of small marten sold in markets in China in the case of SARS; the camel in the Middle East in the case of MERS). Quite astonishingly, some species of bats host very pathogenic agents for humans, such as these coronaviruses, but also viruses like the Ebola virus or the rabies virus, without suffering from any pathology. Recent research has shown that these animals, which have been present on the planet for nearly 65 million years (well before humans, who appeared only 2.5 million years ago), have a particularly efficient immune system, allowing them to protect themselves from these aggressive viruses. In addition, bats have a significant lifespan (40 years for some species, compared to only 2 years for a mouse) and live in groups made up of very large numbers of individuals, which facilitates the transmission of viruses.

© The University of Tours 2025
P. Roingeard, *Journey to the Viral World: Electron Micrographs of Viruses*,
https://doi.org/10.1007/978-3-031-77995-4_57

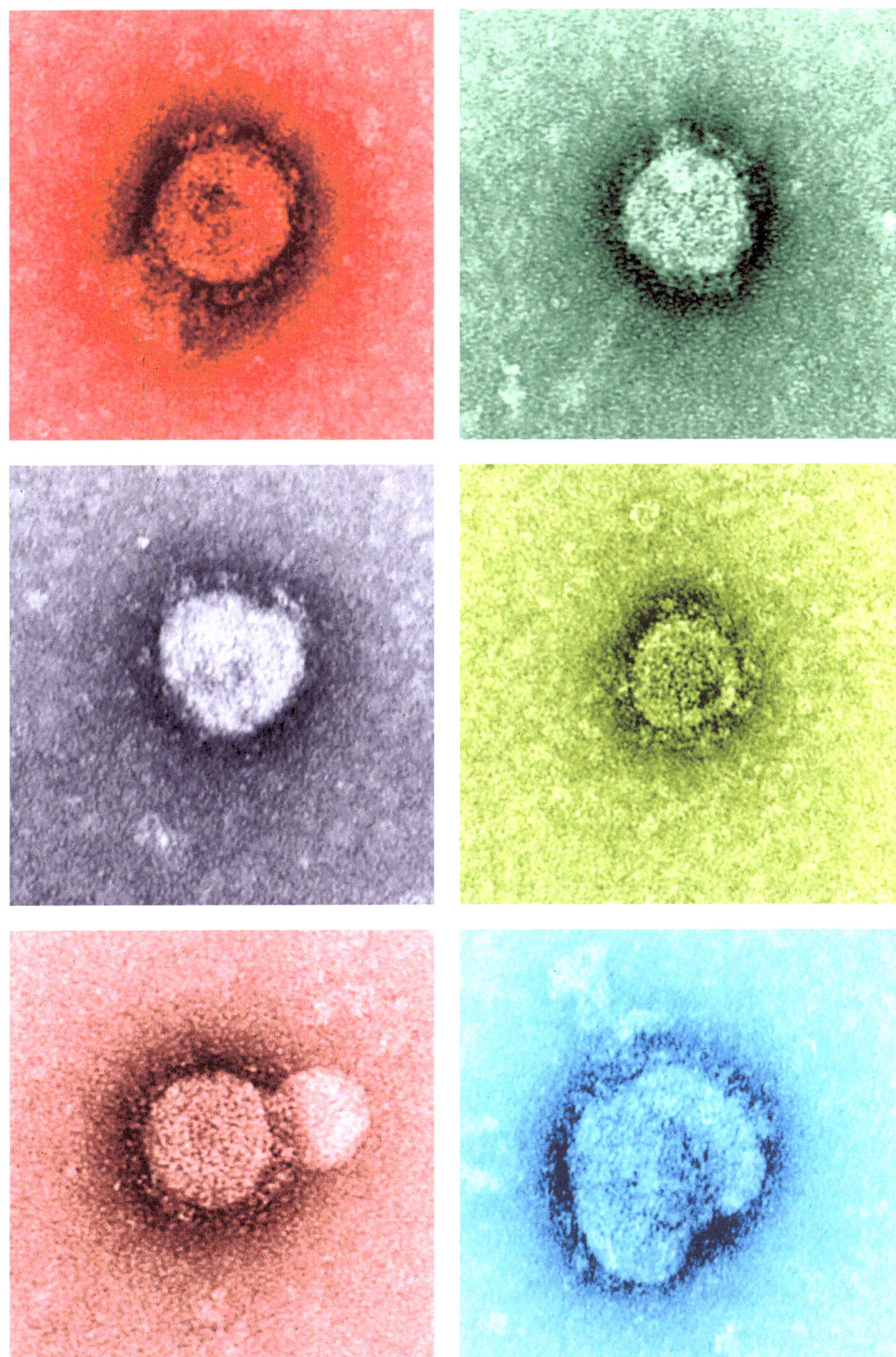

Panel 61. In December 2019, a new form of severe respiratory syndrome, similar to SARS, appeared in the city of Wuhan in China. Even though the precise origin of the transmission is not yet known, it was located in a market in the city, where live wild animals were sold. Again, the virus likely came from a bat and the species barrier was possibly crossed through an intermediate animal which has not been precisely determined. Regardless,

SARS-CoV-2
the cause of
the Covid-19
pandemic

this virus which was named SARS-CoV-2 (and the disease it causes Covid-19 for coronavirus virus disease), turned out to be extremely contagious and highly pathogenic. This photograph shows the SARS-CoV-2 (from a broncho-pulmonary sample of an infected patient) multiplying in cultured cells. The virus (round and dense particles) accumulates in large compartments of the cell.

Initially localized in China during the first weeks of 2020, the virus eventually spread gradually to the entire planet and induced a **pandemic**, i.e. an epidemic that affects a very significant part of the world population, on all 5 continents.

Due to its high contagiousness and **pathogenicity**, many infected people found themselves in respiratory distress, requiring significant needs in medical intensive care units. In order to avoid as much as possible a saturation of these medical units, a majority of countries have implemented measures of population lockdown. Many borders have been closed and transportation means reduced to a strict minimum, leading to a major economic crisis, of global scale.

Even though the WHO declared the end of the international health emergency in May 2023, the virus, which has evolved into different variants, still circulates within the global population. At the end of 2023, the WHO estimated that this **pandemic** had caused about 7 million deaths worldwide. But these figures are certainly well below the reality. Due to a lack of health infrastructure, some countries were late in implementing virus testing and therefore in reporting it as a cause of death. The figure of 27 million deaths, established by the estimate of the excess mortality observed during this period, is thus likely closer to the reality. A century after the **pandemic** of Spanish flu, this **pandemic** linked to SARS-CoV-2 had introduced major social consequences worldwide.

© The University of Tours 2025
P. Roingeard, *Journey to the Viral World: Electron Micrographs of Viruses*,
https://doi.org/10.1007/978-3-031-77995-4_58

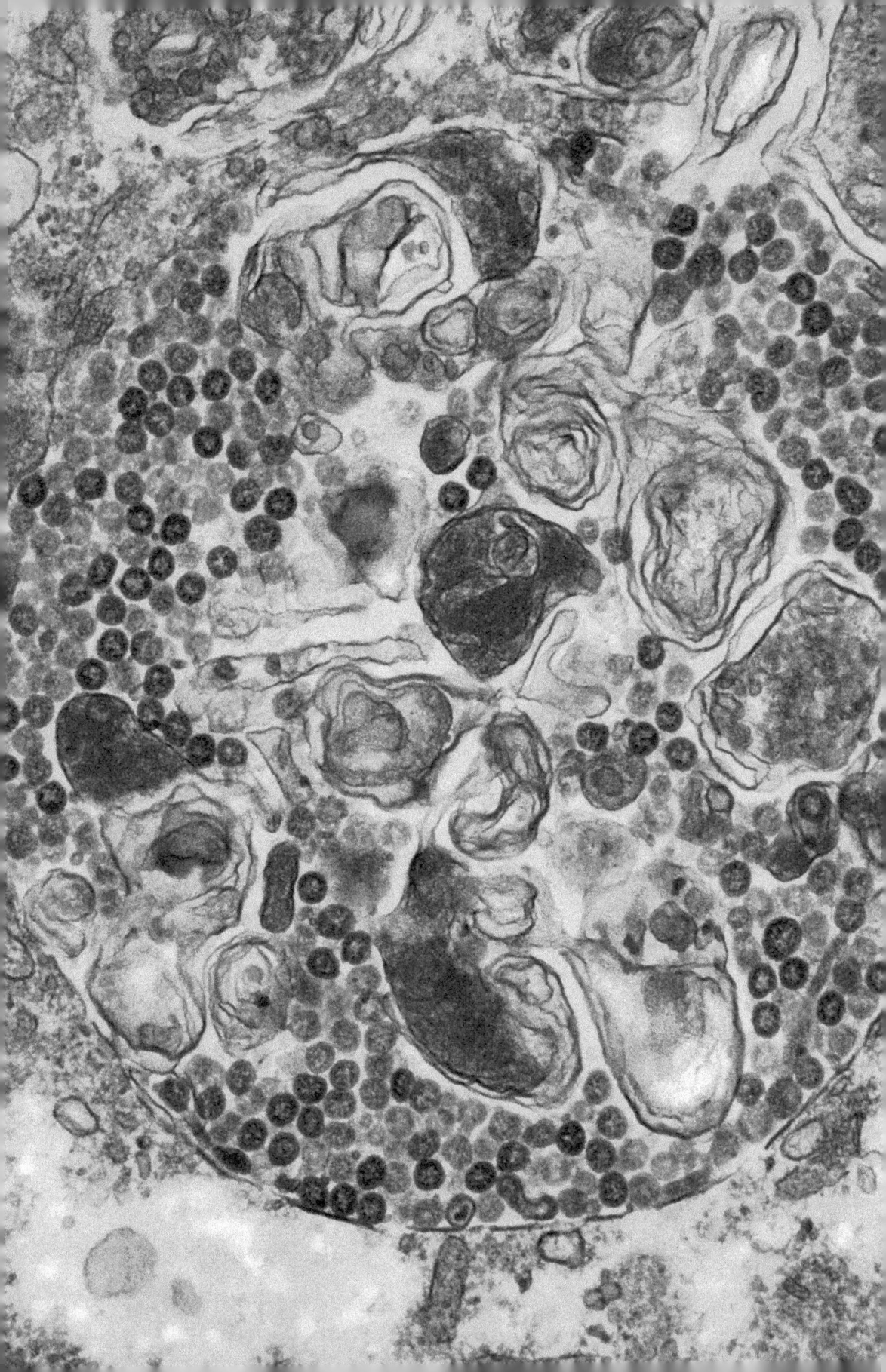

Panel 62. The Covid-19 **pandemic** has generated considerable mobilization of the scientific community (not only virologists) which has quickly produced significant results and strategies to combat the virus. Antiviral treatments and other medical approaches have thus been developed to treat patients infected with SARS-CoV-2, improving survival rates and reducing the severity of symptoms. Scientists have been studying how the virus spreads, deciphering its infectious cycle, as well as modes of airborne and contact transmission, to better understand how to control it.

This transmission electron microscopy image illustrates an example of the study of its infectious cycle, showing how the virus exits an infected cell. The inset bellow shows a virus present in a secretion vesicle of the cell that merges with the cell's plasma membrane, thus releasing the virus into the extracellular environment. The virus makes use of the host cell's physiological secretion mechanisms, diverts and exploits them for its benefit.

SARS-CoV-2
an unprecedented challenge in the scientific community

However, the most spectacular breakthrough in this race against SARS-CoV-2 during the **pandemic** has been the production of effective vaccines that were developed in record time thanks to technological advances, particularly in the use of messenger **RNAs**. The concept is simple: rather than vaccinating with a live attenuated virus, an inactivated virus, or viral protein components produced by genetic engineering, this type of vaccination delivers **RNA** into the muscle cells. The RNA-vaccinated subject will, as a consequence, themselve produce viral proteins thanks to this **RNA**, which stimulate the production of antibodies in the body. The mass production of these "RNA vaccines" has been much faster than conventional vaccines, so that millions of people around the world could have been vaccinated quickly to fight against the spread of the disease.

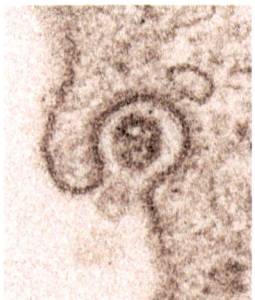

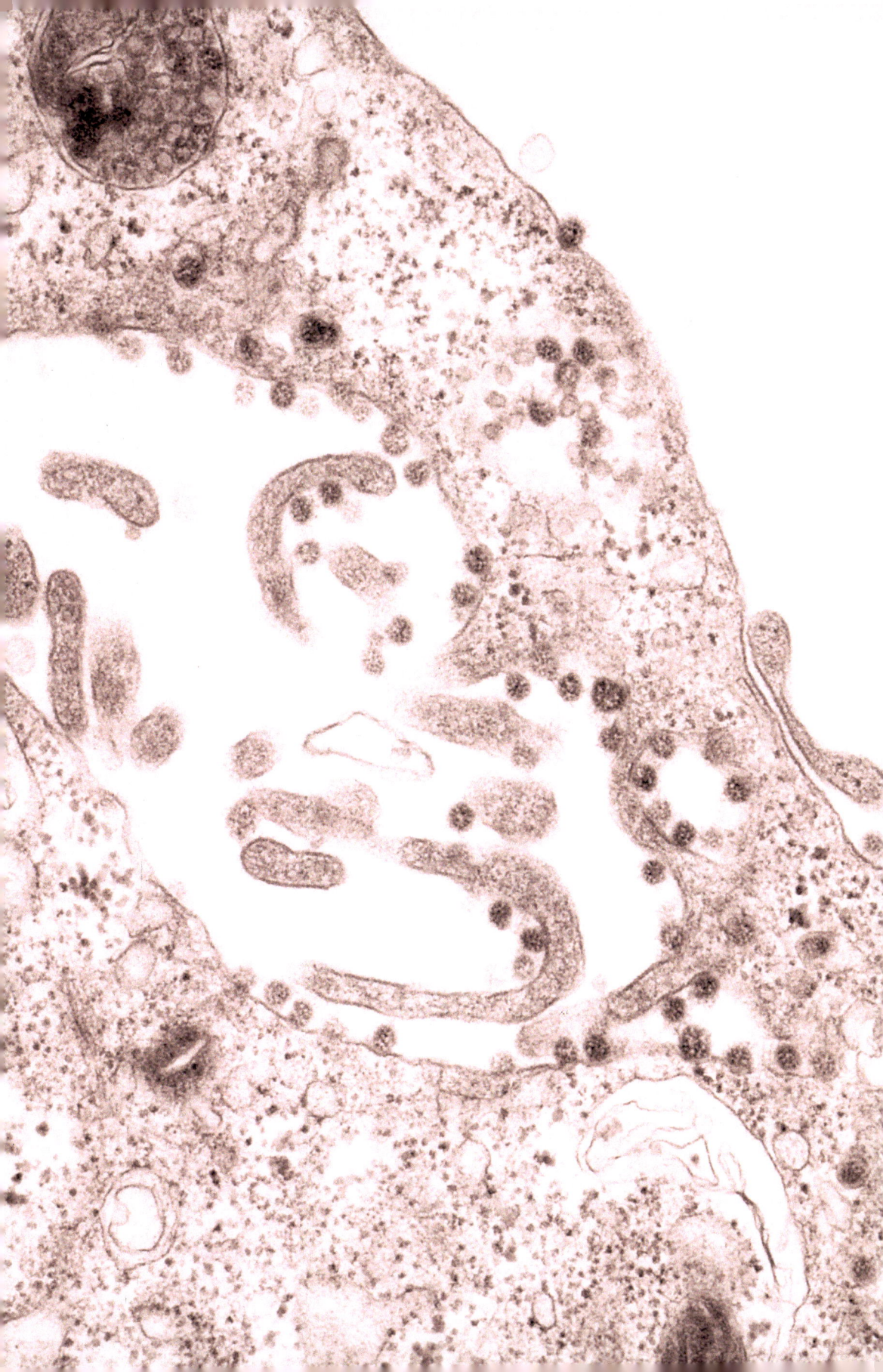

Panel 63. Like other RNA viruses observed on previous panels (West-Nile virus, panel 53; dengue virus panel 54), the SARS-CoV-2 (in green) can also remodel the membranes of the infected cells to induce specific compartments (in pink) in which the virus carries out its genome replication. As with these other viruses, these compartments serve to isolate the replication of the viral genome from sensors triggering the mechanisms of cellular immunity. However, unlike other viruses, these are made up of two membranes, and are called "double membrane vesicles". This photograph clearly shows the two membranes that delimit the inside and outside of these vesicles. The viral RNA replicating inside the vesicles forms a fibrous network that is also clearly visible in this photograph.

Recent research has shown that to form these double membrane vesicles, the virus hijacks the cellular mechanisms of autophagy. Autophagy is a physiological process during which the cell develops a double membrane compartment that encircles cellular structures to degrade and recycle their constituents. SARS-CoV-2 hijacks all the cellular machinery that synthesizes these double membranes to set up these compartments isolating its

Membrane rearrangements induced by the SARS-CoV-2 virus

genome replication. This is once again a remarkable illustration of the ability of viruses to hijack cellular mechanisms for their benefit. Moreover, the recent understanding of the mechanisms leading to the formation of these replication compartments could be essential for the development of new antiviral approaches against SARS-CoV-2 and viruses from other viral families that induce this type of membrane rearrangements.

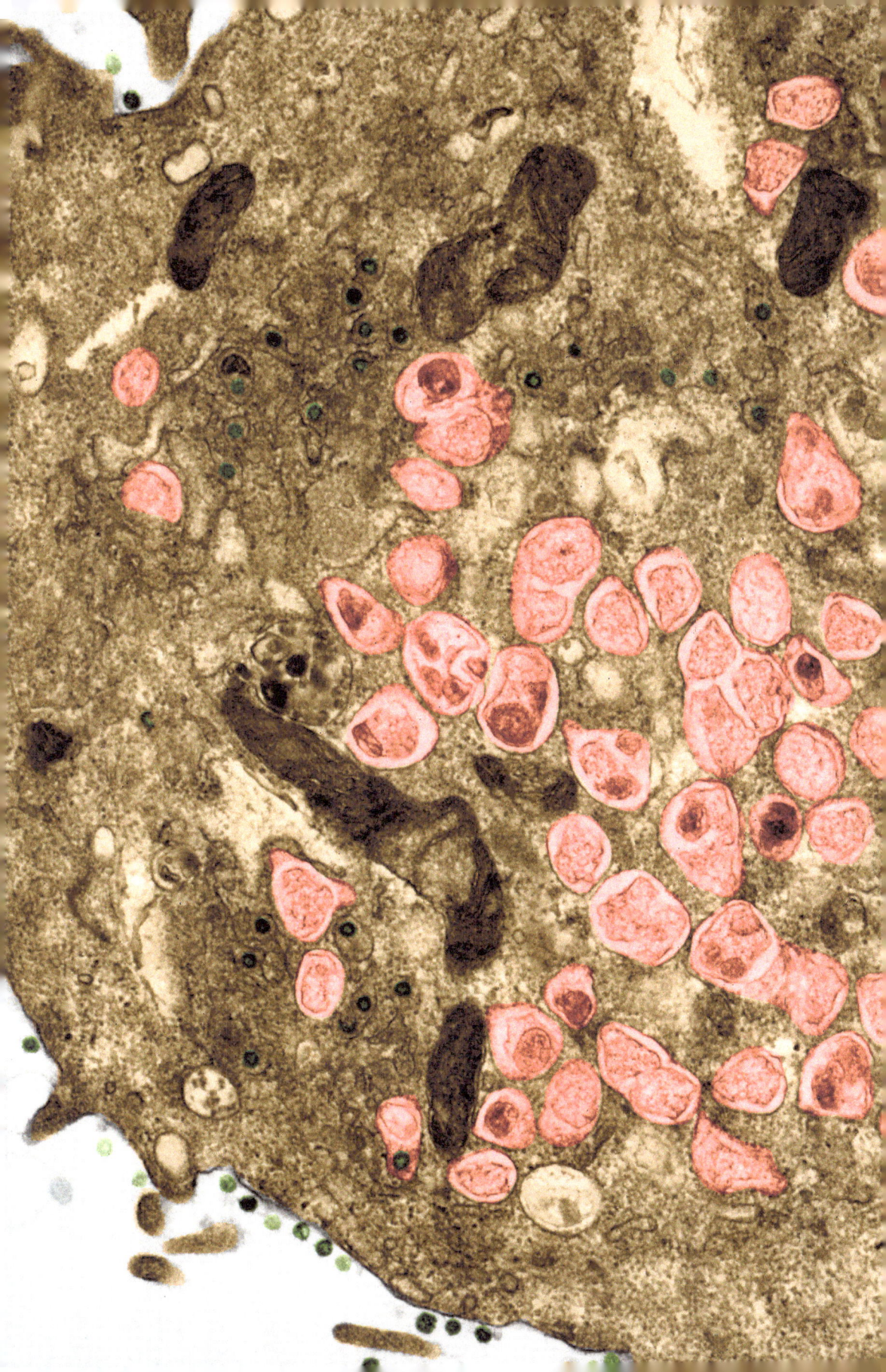

Panel 64. This colorized scanning electron microscopy photograph shows the SARS-CoV-2 (in orange) on the surface of an infected cell (in green). This type of image illustrates that the number of viruses exiting an infected cell at any given time is very high, once again illustrating the extraordinary ability of viruses to multiply within a cell and spread.

The current research on SARS-CoV-2 and Covid-19 mainly focus on monitoring its variants, which is crucial for understanding their impact on transmission, the severity of the disease, and especially the effectiveness of current vaccines in protecting against this evolving virus. Research has also highlighted the long-term effects of the virus in some patients who seem to have eliminated the virus but still have persistent health problems, in a syndrome called "Long Covid". Research on this virus, resulting from the ongoing work of the global scientific community, is constantly evolving, with new studies being regularly published to better understand the virus and the induced disease.

SARS-CoV-2 viruses on a cell surface

The Covid-19 **pandemic** has demonstrated the need to better prepare for the possibility of **pandemics** by investing in strong public health infrastructures, surveillance systems, and rapid response mechanisms. It has also highlighted the importance of international collaboration in dealing with global threats, clearly showing that viruses have no borders.

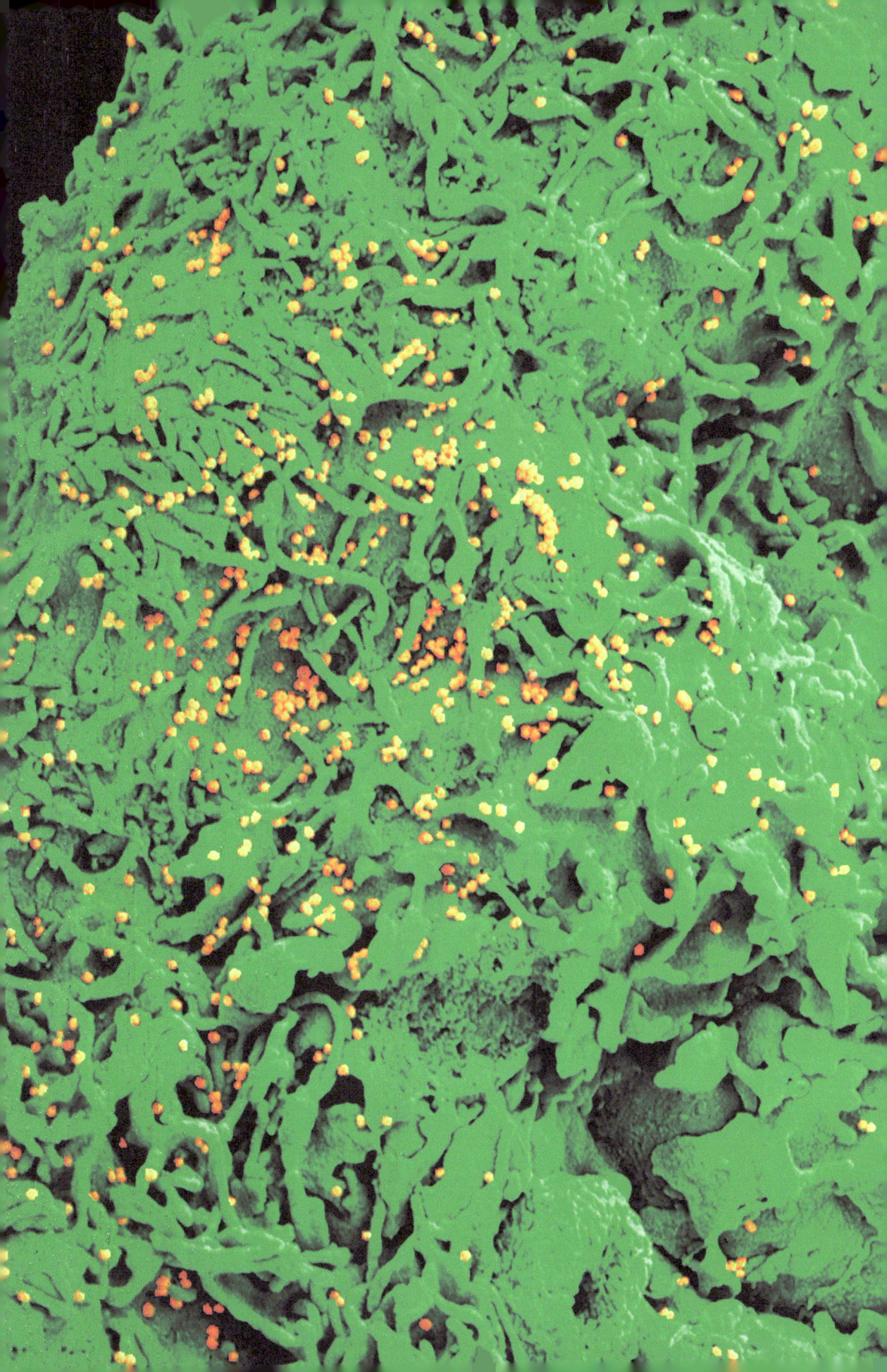

Conclusion

The photographs in this book illustrate most of the viruses that I have "encountered", whether in the context of research or diagnostic activities performed in my laboratory, even if electron microscopy is now much less used than it was 20 years ago to diagnose viral infections. Indeed, electron microscopy played a major role in the discovery of many viruses and in the diagnosis of related diseases, before being gradually replaced by molecular techniques. Such techniques based on the amplification of a viral genome are much more sensitive for detecting minimal amounts of virus in a biological sample. However, to establish a diagnosis with molecular techniques, it is necessary to have an idea of the expected virus in order to specifically amplify its genome. Thus, electron microscopy is still very useful when a virus involved in a particular pathology is totally unknown. In this case, molecular techniques can exclude the role of a given virus, but electron microscopy has the great advantage of being able to visualize and therefore identify the family of the virus involved, thanks to its morphology. This can be done directly from a patient's sample, but most often by contact of this sample with cells in culture in the laboratory. After a few days, sections of these cells are studied under the electron microscope to investigate whether a virus from this sample has multiplied in these cells. Electron microscopy remains also and above all a very important tool for understanding the infectious cycle of viruses in cells, as part of research projects.

Of course, there are other viruses responsible for human diseases that are not presented in this book. It is also important to mention that while scientists know most of the viruses responsible for pathologies, there is a whole community of viruses without any pathogenicity that are totally unknown. Our knowledge is currently progressing rapidly on the bacteria that are present in the human body, which we refer to as the "microbiome". Just like this microbiome, there is undoubtedly a "virome" made up of many non-pathogenic viruses, at least apparently. It will still take years of research before we understand their role. Alongside this, we must exercise vigilance over emerging viruses, which are often known but "re-emerging" viruses. There is no doubt that we will need to continue observing viruses with electron microscopes to better understand them.

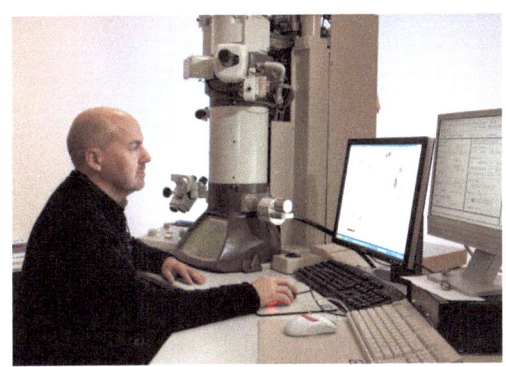

© The University of Tours 2025
P. Roingeard, *Journey to the Viral World: Electron Micrographs of Viruses*,
https://doi.org/10.1007/978-3-031-77995-4_59

The 4th Cover Summary

Viruses are fascinating in many aspects. They have been responsible for major epidemics that have marked the history of humanity such as smallpox, Spanish flu, AIDS, the recent epidemic linked to the SARS-CoV-2 coronavirus, and others may be yet to come. They are so small that extremely powerful microscopes, electron microscopes, are necessary to observe them. Unable to multiply on their own, they have found all possible tricks to enter a cell and exploit the cellular machinery for their benefit, in order to multiply on a large scale and spread.

This book is mainly a picture book. Electron microscopy images, associated to short and simple captions, allow a better understanding of the world of viruses: their morphology, their journey in the cell they infect, the diseases they can generate. The reader will be able to visualize the cause of infectious diseases such as measles, chickenpox, mumps, rubella, flu, rabies, viral hepatitis, AIDS, and many others... Despite the fact that some viruses mentioned in this book are the source of particularly serious diseases, one can be captivated by the images and the aesthetics of certain viral structures.

Author Presentation

Philippe Roingeard is a professor of cell biology at the School of Medicine of the University of Tours and a hospital practitioner at the University Hospital (CHU) of Tours, in France. He heads the IBiSA electron microscopy platform of the University and CHU of Tours. He is a member of the INSERM U1259 research team "Morphogenesis and Antigenicity of HIV, hepatitis viruses and emerging viruses", which he led between 2006 and 2023. He has received several awards for his research on viruses: National Academy of Medicine (2014), Foundation of France (2015), Grand Prize of the National Academy of Pharmacy (2018). He is corresponding member of the National Academy of Medicine since 2022.

Catalog Summary

Viruses have been responsible for major epidemics that have marked the history of humanity such as smallpox, Spanish flu, AIDS, the recent epidemic linked to the SARS-CoV-2 coronavirus, and others may be yet to come. They are so small that they can only be observed with electron microscopes. Unable to multiply on their own, they have found extraordinary tricks to enter a cell and exploit all the cellular machinery for their benefit. The electron microscopy images of this book, associated to short captions, allow a better understanding of the world of viruses. Despite the fact that some of these viruses are the source of serious diseases, one can be captivated by the aesthetics of certain images.